Lecture Notes in Computer Science 16184

Founding Editors

Gerhard Goos
Juris Hartmanis

The series Lecture Notes in Computer Science (LNCS), including its subseries Lecture Notes in Artificial Intelligence (LNAI) and Lecture Notes in Bioinformatics (LNBI), has established itself as a medium for the publication of new developments in computer science and information technology research, teaching, and education.

LNCS enjoys close cooperation with the computer science R & D community, the series counts many renowned academics among its volume editors and paper authors, and collaborates with prestigious societies. Its mission is to serve this international community by providing an invaluable service, mainly focused on the publication of conference and workshop proceedings and postproceedings. LNCS commenced publication in 1973.

Benjamin Hou · Tejas S. Mathai
Editors

Learning with Longitudinal Medical Images and Data

First International Workshop, LMID 2025
Held in Conjunction with MICCAI 2025
Daejeon, South Korea, September 27, 2025
Proceedings

Editors
Benjamin Hou
National Institutes of Health
Bethesda, MD, USA

Tejas S. Mathai
National Institutes of Health
Bethesda, MD, USA

ISSN 0302-9743 ISSN 1611-3349 (electronic)
Lecture Notes in Computer Science
ISBN 978-3-032-16127-7 ISBN 978-3-032-16128-4 (eBook)
https://doi.org/10.1007/978-3-032-16128-4

This Springer imprint is published by the registered company Springer Nature Switzerland AG
The registered company address is: Gewerbestrasse 11, 6330 Cham, Switzerland

Preface

On behalf of the organizing committee, we welcome you to the Workshop on Learning with Longitudinal Medical Images and Data (LMID 2025), held in-person on September 27 2025 at the 2025 Medical Image Computing and Computer-Assisted Intervention (MICCAI) conference in Daejeon, South Korea. This was the 3rd edition of the workshop, which has been organized since 2023. It was organized with the combined efforts of researchers at the National Institutes of Health (NIH), Radboudumc, Fraunhofer MEVIS, and the Icahn School of Medicine at Mount Sinai.

The goal of the workshop was to foster new ideas on longitudinal medical data analysis, tracking, and modeling disease progression with imaging and/or multimodal data. Medical data, including imaging, is routinely employed to track the progression of disease or to assess response to treatment. Recent developments in artificial intelligence (AI) and machine learning have shown promise in automating or improving parts of the clinical workflow, from being able to find lesions in various organs, to classification and diagnosis of diseases. Unfortunately, despite the key role of serial imaging in the clinical workflow, developing AI systems that can track or model disease progression by learning from and exploiting longitudinal imaging has not received much attention until recently. This workshop aimed to bring together researchers and clinicians to tackle the important challenges facing longitudinal data analysis. This year's workshop focused on brain imaging, lesion tracking and interval change assessment, glaucoma prognosis, and knee osteoarthritis diagnosis.

The LMID workshop attracted 14 submissions this year : 13 full-length papers (10 pages long) and 1 short paper were submitted. Each full paper underwent rigorous evaluation through a double-blind peer review process. Two independent reviewers were involved in the review process for each paper, and a meta reviewer from the program committee assessed the comments to recommend either acceptance or rejection of the paper. Based on review scores and committee feedback, 12 full-length papers were selected for presentation and inclusion in the workshop proceedings. The short paper was accepted for presentation at the workshop but not included in the final proceedings. We would like to sincerely thank the reviewers and committee members for their dedicated efforts to review the papers and ensure the workshop's high academic quality standards.

September 2025

Tejas S. Mathai
Benjamin Hou

Organization

Proceedings Chairs

Tejas S. Mathai	National Institutes of Health, USA
Benjamin Hou	National Institutes of Health, USA

Workshop Chairs

Tejas S. Mathai	National Institutes of Health, USA
Alessa Herring	Radboudumc, The Netherlands
Pritam Mukherjee	National Institutes of Health, USA
Benjamin Hou	National Institutes of Health, USA
Xinya Wang	National Institutes of Health, USA
Boah Kim	Sungkyunkwan University, South Korea
Yan Zhuang	Icahn School of Medicine at Mount Sinai, USA

Contents

SSL-AD: Spatiotemporal Self-supervised Learning for Generalizability and Adaptability Across Alzheimer's Prediction Tasks and Datasets

Emily Kaczmarek[1,2](✉), Justin Szeto[1,2], Brennan Nichyporuk[1,2], and Tal Arbel[1,2]

[1] Centre for Intelligent Machines, McGill University, Montreal, Canada
[2] Mila – Quebec AI Institute, Montreal, Canada
emily.kaczmarek@mail.mcgill.ca

Abstract. Alzheimer's disease is a progressive, neurodegenerative disorder that causes memory loss and cognitive decline. While there has been extensive research in applying deep learning models to Alzheimer's prediction tasks, these models remain limited by lack of available labeled data, poor generalization across datasets and tasks, and the inability to leverage temporal patterns across varying numbers of input scans. In this study, we adapt state-of-the-art temporal self-supervised learning (SSL) approaches for 3D brain MRI analysis and add novel extensions designed to handle variable-length inputs and learn robust spatial features, resulting in three distinct pre-training strategies. We aggregate four publicly available datasets comprising 3,161 patients for pre-training, and show the performance of our models across multiple Alzheimer's prediction tasks including diagnosis classification, conversion detection, and future conversion prediction. Importantly, our SSL model implemented with temporal order prediction and contrastive learning outperforms supervised learning on six out of seven downstream tasks. By fine-tuning the same pre-trained backbone with one, two, or three input images, we demonstrate that our model is adaptable to varying numbers of input images and time intervals, as well as generalizable across tasks, highlighting its robust clinical potential. We release our code and model publicly at https://github.com/emilykaczmarek/SSL-AD.

Keywords: Temporal Self-Supervised Learning · Alzheimer's Disease · Representation Learning · 3D Brain MRI

1 Introduction

Many chronic and neurodegenerative disorders progress gradually over time, making longitudinal prediction an important but challenging clinical goal. Alzheimer's Disease (AD) is an incurable, neurodegenerative disorder characterized by progressive memory loss, cognitive decline, and behavioural changes [13]. Typically, AD develops gradually, with Cognitively Normal (CN) individuals first progressing to Mild Cognitive Impairment (MCI), and potentially further transitioning to AD. AD motivates a range of clinically important spatial

B. Hou and T. S. Mathai (Eds.): LMID 2025, LNCS 16184, pp. 1–12, 2026.
https://doi.org/10.1007/978-3-032-16128-4_1

and temporal prediction tasks, including (i) current disease status, (ii) detecting the transition to MCI/AD, and (iii) predicting future transition to MCI/AD. Deep learning has been extensively applied to these tasks, with multiple review papers summarizing the progress in this field [6,7,12,22]. The majority of existing studies use supervised learning, which requires labeled data and separate models trained for each specific task. This severely limits both the amount of training data available, and the generalizability across tasks and/or population changes. In addition, most approaches require a fixed number of scans across time per patient. This further reduces the training set by excluding patients with fewer scans, and prevents learning temporal relationships that generalize across patients with differing numbers of visits. While supervised learning has shown strong results for Alzheimer's prediction tasks, it remains limited by labeled data availability, poor representation generalizability, and lack of adaptability to varying numbers of images per patient.

Recently, in machine learning, self-supervised learning (SSL) has emerged as a popular alternative to supervised learning [5]. In SSL, general-purpose representations are learned by deriving tasks using the input data alone (i.e., without external labels). This allows models to be trained on larger, unlabeled datasets (often aggregated from many sources), and fine-tuned using limited labels on numerous downstream tasks. In addition to spatial SSL models, several temporal SSL approaches have been introduced, including temporal order verification (predicting if a sequence of images or video clips is in the correct order) [14] and temporal order prediction (predicting the exact permutation of images/clips) [23]. These models are well-suited to analyzing 3D brain MRI of neurodegenerative diseases; given the similarity of brain images across time, the model is encouraged to focus on subtle but important temporal changes. While a number of SSL models have been developed for Alzheimer's prediction, they remain limited by (1) lack of larger, aggregated pre-training datasets [8,20,24], (2) omission of temporal learning objectives [8,9,24], and (3) their inability to train with varying numbers of input images [8,9,20,24]. There remains a lack of models that truly exploit temporal self-supervision to improve adaptability and performance across Alzheimer's prediction tasks.

In this paper, we present the first fully adaptable, generalizable self-supervised framework for analyzing 3D longitudinal brain MRI for Alzheimer's prediction tasks. We adapt state-of-the-art temporal SSL approaches (originally developed for natural videos) to longitudinal 3D brain MRI analysis and add novel extensions for adaptability to varying numbers of input images and improved spatial features, resulting in three novel pre-training strategies. Specifically, we first implement temporal order verification, followed by temporal order prediction with a novel approach to support different numbers of input images. Our third model combines the temporal order prediction approach with contrastive learning to improve both spatial and temporal representations. All SSL models are trained using the aggregation of four publicly available datasets (HABS-HD [15], MCSA [17], NIFD [18], ADNI [16]) comprising a total of 3,161 patients. We evaluate our models on a subset of held-out test data of CN, MCI and AD patients. Our models are adaptable to being fine-tuned on one to four images acquired at varying time intervals, showing improved performance with longer image sequences and enabling its application in real-world clinical settings. Further, our model is generalizable across multiple tasks with simple fine-tuning, shown through the classification of CN, MCI, and AD subjects, as well as the detection and future prediction of patients converting from CN to MCI, and from MCI to AD. Through extensive experiments, we find that the temporal order prediction model with contrastive learning consistently has the best performance, and outperforms supervised learning on six of the seven downstream tasks. Notably, this pre-training approach can be applied to any pre-existing model architecture developed for Alzheimer's prediction tasks. We release all code and models for public use to support reproducibility and enable further development.

2 Methodology

The objective of our work is to develop a temporal self-supervised model that is able to perform challenging detection and prediction tasks that rely on longitudinal information. We present an overview of our three temporal self-supervised models for 3D brain MRI analysis, adapted from state-of-the-art SSL methods for natural videos. These models are designed to correctly predict the temporal order of a 3D brain MRI sequence; given the similarity between scans over time, a model must identify key small changes to solve the ordering task. First, we begin with an architecture based on Shuffle and Learn [14], a popular method in the context of computer vision videos. During training, we create a binary prediction task to determine if a sequence of 3D brain MRI scans across time is correctly ordered (*temporal order verification*), and call this model SSL-TOV. Second, we extend the pre-training task to predicting the exact permutation of the sequence (*temporal order prediction* [23]), with a novel extension to support classification of two, three, and four MRI input permutations (SSL-TOP). Third, we combine the temporal order prediction approach with contrastive learning to simultaneously learn temporal and spatial features (SSL-TOPC, *temporal order prediction contrastive*). All architectures are visualized in Fig. 1.

2.1 SSL-Temporal Order Verification (SSL-TOV)

The first learning paradigm we examine for longitudinal image analysis is temporal order verification. Given a patient with a sequence of N temporally ordered 3D volumes, $S = (\mathbf{x}_{t_1}, \mathbf{x}_{t_2}, \ldots, \mathbf{x}_{t_N})$, where each $\mathbf{x}_{t_n} \in \mathbb{R}^{C\times D\times H\times W}$ represents a scan acquired at time t_n, we randomly permute the sequence and assign a label where $y = 1$ if $t_1 < t_2 < \cdots < t_N$, and $y = 0$ otherwise. In other words, the sample is labeled positive if all images appear in the correct chronological order, and negative if any image is out of order. After permutation, to account for variability in the number of scans each patient has, we pad sequences to a fixed length T (if $N < T$) by appending zero-filled volumes. Each image in the sequence is then individually encoded by a CNN backbone, $f_\theta(\cdot)$, and the T

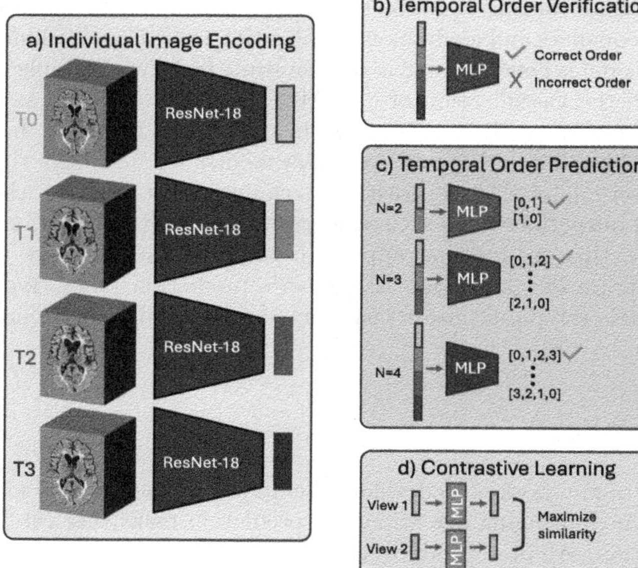

Fig. 1. Overview of all three temporal SSL approaches. (a) For each method, up to four 3D brain MRI scans (1–2.5 years apart) are encoded separately by a ResNet-18 backbone. (b) Temporal Order Verification (SSL-TOV) is performed by concatenating image representations and using a multi-layer perceptron (MLP) to predict if the images are in the correct order. If a patient has less than four images, their sequence is padded with zero-filled volumes. (c) Temporal Order Prediction (SSL-TOP) takes the concatenated image representations and predicts the exact permutation of the sequence. Here, three different classifiers are used to predict the permutation for two, three, and four images (no padding is used). (d) The final temporal SSL model (SSL-TOPC) combines temporal order prediction with contrastive learning, where two different views of a sequence are created through augmentations. Images are encoded individually, and the two augmented representations of a selected image in a sequence are projected through an MLP. The similarity of the projections is maximized through the NT-Xent loss.

image representations are concatenated before being passed to multi-layer perceptron classifier, $g_\phi(\cdot)$. The model is trained to minimize the standard binary cross-entropy loss (shown for a single patient):

$$\ell_{\text{BCE}} = -\left[y \log \hat{y} + (1 - y) \log(1 - \hat{y})\right], \tag{1}$$

where $y \in \{0, 1\}$ is the ground truth label indicating whether the sequence is in correct order, and $\hat{y}$ is the model's prediction.

2.2 SSL-Temporal Order Prediction (SSL-TOP)

Following temporal order verification, we consider a more challenging self-supervised task: temporal order prediction. Instead of performing binary classification to predict if the sequence is in order, the model is trained to predict the exact permutation of the image sequence. This formulation increases the task difficulty and encourages the model to focus on specific and subtle differences between all timepoints.

More precisely, for a sequence S of N images, there are $N!$ possible permutations, making the task an $N!$-way classification problem. Since the number of images per patient varies, we develop a novel model modification to enable permutation classification. Specifically, after using the same shared backbone encoder $f_\theta(\cdot)$ as in temporal order verification, the concatenated representations are passed to a permutation classifier corresponding to the sequence length, $g_\phi^{(N)}(\cdot)$. In practice, this means that $g_\phi^{(2)}(\cdot)$ will have $2! = 2$ classes, $g_\phi^{(3)}(\cdot)$ will have $3! = 6$ classes, and so on. The shared backbone of the model ensures that general temporal features are learned, while the permutation classifiers encourage the model to learn specific differences between timepoints in sequences of varying lengths. For a single patient with N images in their sequence, resulting in $N!$ classes, the categorical cross-entropy loss is:

$$\ell_{\mathrm{CE}}^{N} = -\sum_{k=1}^{N!} y_k \log \hat{y}_k,$$

where y_k is the one-hot encoded ground truth permutation and $\hat{y}_k$ is the predicted probability for class k. This formulation no longer requires image padding, as each sequence length N has its own dedicated classifier. In addition, instead of passing every possible sequence to its corresponding classifier each epoch, resulting in the summation of multiple losses ℓ^N (one loss per classifier head $g_\phi^{(N)}(\cdot)$), we simply sample and train on sequences with the same randomly chosen N for a given epoch (i.e., one epoch focuses on sequences with $N = 2$, the next focuses on $N = 4$, and so on).

2.3 SSL-Temporal Order Prediction with Single-Timepoint Contrastive Learning (SSL-TOPC)

One of the most common and successful forms of self-supervision is contrastive learning. In contrastive learning, two different views of an image are created through augmentations, and the objective is to maximize the similarity between projections of the two augmented representations. To improve spatial feature learning and develop a stronger backbone encoder $f_\theta(\cdot)$, we add single-timepoint contrastive learning to our temporal order prediction model.

Formally, each temporal brain MRI sequence, S, is randomly augmented to create two views, $S_i = a_i(S), S_j = a_j(S)$, where a_i, a_j are independent random augmentations (discussed in Sect. 3.2). For a given view, the same augmentations are applied consistently to all images within the sequence. After encoding each image in each sequence (without applying any permutation), the representations of a single randomly selected image x_{t_r} in both augmented sequences $\mathbf{h}_{t_r,i} = f_\theta(a_i(x_{t_r})), \mathbf{h}_{t_r,j} = f_\theta(a_j(x_{t_r}))$ are projected through a multi-layer perceptron $\mathbf{z}_{t_r,i} = p_\psi(\mathbf{h}_{t_r,i}), \mathbf{z}_{t_r,j} = p_\psi(\mathbf{h}_{t_r,j})$. Only a single image is chosen to reduce the computational burden of applying contrastive learning to all images in a sequence, and we select the same image x_{t_r} from both independently augmented sequences. The projections are pushed together to maximize similarity, and pulled apart from all other selected images in a batch using the NT-Xent loss [4] (we remove the t_r subscript for clarity):

$$\ell_{i,j} = -\log \frac{\exp(\text{sim}(\mathbf{z}_i, \mathbf{z}_j)/\tau)}{\sum_{k=1}^{2B} \mathbb{1}_{[k \neq i]} \exp(\text{sim}(\mathbf{z}_i, \mathbf{z}_k)/\tau)}, \tag{2}$$

where B is the batch size and τ is the temperature parameter. The total NT-Xent loss for a given patient is calculated using both z_i and z_j as the reference (i.e., the representation used as the comparison in the denominator), summing the loss in both directions: $\ell_{\text{NT-Xent}} = \frac{1}{2}[\ell_{i,j}(\mathbf{z}_i, \mathbf{z}_j) + \ell_{j,i}(\mathbf{z}_j, \mathbf{z}_i)]$. Subsequently, the image representations from one of the augmented sequences are randomly permuted, concatenated, and input to the correct permutation prediction head as described in Sect. 2.2. The final loss is the sum of the contrastive and classification losses:

$$\ell = \ell_{\text{NT-Xent}} + \ell_{\text{CE}} \tag{3}$$

Since only a single image in each sequence is used for the contrastive learning loss, we label this as *single-timepoint contrastive learning.*

3 Experiments and Implementation Details

We pre-train our temporal SSL models on 3D brain MRI aggregated data from four publicly available longitudinal datasets focused on healthy aging and dementia. Each dataset is briefly described below, including our evaluation protocol using ADNI. We evaluate the performance of our models on three specific types of tasks: (1) classification of stable (i.e., constant disease states across input images) CN, MCI, and AD patients using one, two, and three images (labels are determined using three consecutive clinical timepoints, as outlined below in Sect. 3.1); (2) detection of conversion from one state to another using two images (e.g., stable CN subjects versus those who converted to MCI); (3) prediction of future conversion to MCI or AD (from CN and MCI, respectively) using a single image. After the dataset and evaluation protocol, we describe our preprocessing and implementation details for the SSL models and supervised baseline.

3.1 Datasets

We use four publicly available datasets for pre-training the SSL models. The *Health and Aging Brain Study – Health Disparities (HABS-HD)* [15] and *Mayo Clinic Study of Aging (MCSA)* [17] datasets contain 4,231 and 1,802 patients, respectively, representing CN, MCI and AD patients. The *Neuroimaging in Frontotemporal Dementia (NIFD)* [18] dataset contains 346 healthy controls and patients with frontotemporal or other forms of dementia.

We demonstrate the ability of our model to adapt to different numbers of input images and downstream tasks using the *Alzheimer's Disease Neuroimaging Initiative (ADNI)* [16] dataset, which includes CN, MCI, and AD patients. We divide ADNI into training (2,074 patients), validation (471 patients), and test (483 patients) splits prior to pre-training. The training split is used, in aggregation with the previous three datasets, for SSL pre-training. Since pre-training focuses on longitudinal image analysis, we omit all patients that only have a single image, creating an SSL pre-training dataset consisting of 3,161 patients. During SSL fine-tuning and supervised baseline training, subsets of this training data are selected based on available labels. The validation and test data are completely held-out from pre-training and used only for downstream evaluation.

For consistency across tasks, we maintain these ADNI splits and sample task-specific train/validation/test sets within them. For the classification task, we extract triplets with images that are 1–2.5 years apart with a constant label (e.g., MCI-MCI-MCI). Next, we remove the last image from the triplet to create a pair (e.g., MCI-MCI) and the two last images to create a single timepoint (e.g., MCI). This ensures that the number of patients remains constant across experiments, meaning any performance differences are based on input sequence length as opposed to dataset size. In total, there are 278 patients for training (611 triplets/pairs/images), 80 patients in validation, and 90 patients in test. We perform similar sampling for our conversion detection and prediction tasks: all possible pairs of stable and converting images with a gap of 1–2.5 years are used for training the conversion detection task (e.g., MCI-AD), and the first image in each pair is used for training the future prediction of conversion (e.g., MCI). For CN to MCI conversion, we use 373 patients for training (1,787 pairs/images), 111 for validation, and 124 for test, and for MCI to AD, we use 516 patients for training (2,483 pairs/images), 157 for validation, and 155 for test.

3.2 Data Preprocessing

All datasets are preprocessed using the same standardized pipeline implemented through TurboPrep[1]. This publicly available repository has combined common (and fast) 3D brain MRI preprocessing techniques for easy implementation. Briefly, we perform N4 bias-field correction to harmonize and remove inconsistencies from scanners [21], skull stripping to remove non-brain regions [11], linear registration to align scans to a standard size and shape [1], segmentation to identify specific anatomical regions [2], and intensity normalization for consistent intensities across subjects [19]. Images are resampled to a resolution of $1 \times 1 \times 1$ mm^3 ($193 \times 229 \times 193$ voxels), permuted such that the image depth is first dimension, and cropped to ($150 \times 192 \times 192$).

During pre-training, we apply augmentations to images across all SSL models using the Medical Open Network for Artificial Intelligence (MONAI [3]). These augmentations consist of an affine transformation (rotation up to $\pm$ 0.34 rad,

[1] https://github.com/LemuelPuglisi/turboprep.

translation ± 15 voxels, scaling in [0.9, 1.3]) applied across all timepoints, followed by independent Gaussian smoothing (sigma in [0.25, 1.5]) and noise (standard deviation in [0.05, 0.09]) in attempt to remove differences between scanners across time. Each augmentation is applied with 0.5 probability. Lastly, we normalize the images by subtracting the overall training set standard deviation. For downstream training or fine-tuning, augmentations include a random spatial crop with minimum size of $(90 \times 115 \times 115)$ and resize to $(150 \times 192 \times 192)$, random horizontal flip (probability=0.5), random affine transformation with rotation ±0.1 rad, scaling ± 15%, and translation ± 5 voxels (probability=0.7), random intensity shift with offset 0.1 (probability=0.5), and random Gaussian noise added with mean of 0, standard deviation of 0.1 (probability=0.2).

3.3 Model Details

For SSL pre-training, we acquire pairs (3,161 patients), triplets (1,114), and quadruplets (445) of images that are 1–2.5 years apart. While additional images may provide increased temporal information, they create increased challenges with missing data, as well as overcomplicated permutation tasks (i.e., five images yields 120 possible classes). The selection of up to four images that are 1–2.5 years apart allows a period of 1–10 years to be analyzed. For patients that have more than a single image/pair/triplet available, we sample a single data input each epoch to ensure patients are equally represented throughout training.

Each SSL model is implemented with a ResNet-18 backbone and a two-layer multi-layer perceptron classifier of size (512, 256). The projection dimension for the single-timepoint contrastive projection is 128. For downstream tasks, we condition all supervised and SSL models on the time gap between scans when training on two or more images. The final layer of the pre-trained SSL classifier (from binary or permutation prediction) is replaced with a linear layer with the desired number of classes. When training with a single image downstream, only the pre-trained backbone is used, and a two-layer multi-layer perceptron is trained from scratch. For our supervised baseline comparison, we use the same ResNet-18 [10] backbone (with additional channels to encode multiple timepoints as needed), and a two-layer multi-layer perceptron classifier. This controlled setup isolates the improvement due to the proposed SSL pre-training approach, which is architecture-agnostic and can be applied to any backbone.

4 Results

We assess the utility of our pre-training regimes against a supervised ResNet-18 model. First, to demonstrate the ability of our model to be fine-tuned on varying numbers of input images (and the importance of this ability), we classify stable Alzheimer's patients against stable MCI and CN subjects. Next, using two MRI scans, we detect patients who have converted from one disease state to another. Lastly, using a single MRI scan as input, we predict conversion to another disease

Table 1. AUC performance across tasks (higher is better) for the supervised ResNet-18, and SSL-TOV, SSL-TOP, and SSL-TOPC models. For classification of stable CN, MCI, and AD, we use one-vs-rest AUC. Bold indicates best in each row, underline indicates second best. The SSL-TOPC has the highest performance most consistently, outperforming the supervised model on six of the seven tasks. SL: Supervised Learning; CN: Cognitively Normal; MCI: Mild Cognitive Impairment; AD: Alzheimer's Disease.

Task	SL ResNet-18	SSL-TOV	SSL-TOP	SSL-TOPC
Classification of Stable CN, MCI, and AD Patients				
1 MRI Scan	0.673±0.041	<u>0.683±0.037</u>	0.680±0.048	**0.712±0.057**
2 MRI Scans	<u>0.767±0.015</u>	0.748±0.019	0.745±0.021	**0.789±0.028**
3 MRI Scans	0.780±0.020	<u>0.848±0.011</u>	**0.872±0.030**	0.838±0.004
Conversion Detection with Two Images				
CN → MCI (2 MRI)	<u>0.639±0.042</u>	0.634±0.037	0.610±0.128	**0.650±0.057**
MCI → AD (2 MRI)	<u>0.769±0.034</u>	0.722±0.028	**0.778±0.021**	0.719±0.024
Future Conversion Prediction with a Single Image				
CN → MCI (1 MRI)	<u>0.680±0.018</u>	0.653±0.056	0.642±0.038	**0.697±0.053**
MCI → AD (1 MRI)	0.633±0.043	<u>0.686±0.023</u>	**0.693±0.021**	0.667±0.045

state 1–2.5 years in the future, showing the benefit of pre-training on longitudinal objectives to learn generalizable representations with temporal information.

Table 1 shows AUC performance for seven different Alzheimer's prediction tasks. We repeat each downstream task three times and report the mean and standard deviation. Overall, the SSL-TOPC model shows superior performance for six out of the seven tasks compared to supervised learning. Importantly, we notice that this model particularly excels when training on a single image. For example, when classifying stable CN, MCI, and AD patients, and predicting future conversion of CN to MCI and MCI to AD, the SSL-TOPC has a 4%, 1%, and 3% increase over supervised baselines (respectively), and higher performance than the other SSL models for two of these tasks. This highlights the benefit of extracting strong spatial features through contrastive learning, particularly when temporal information is unavailable in downstream applications. The SSL-TOV and SSL-TOP models do not outperform the supervised ResNet-18 as often as SSL-TOPC, further demonstrating the importance of combining temporal and spatial SSL learning objectives. However, we note that SSL-TOV and SSL-TOP do outperform SSL-TOPC for some tasks, suggesting that future work should focus on identifying the cases where each SSL model is most effective.

4.1 The Effect of Multiple Timepoints

The first three rows of Table 1 show the performance of classifying stable CN, MCI, and AD patients, using one, two, or three input images. All models (both supervised and SSL) perform better as the number of input scans increases. The SSL-TOPC outperforms supervised learning on all tasks; this highlights the importance of developing SSL pre-training approaches that can adapt to varying numbers of input images, leveraging all available information for improved prediction across downstream tasks with different data availability.

4.2 Conversion Detection

The next task is conversion detection (rows 4 and 5 in Table 1), where two images are used to classify if conversion has occurred. Interestingly, the supervised ResNet-18 performs much better than the SSL-TOPC model, and almost as well as the SSL-TOP model at detecting the conversion from MCI to AD. Alzheimer's disease has significant visible brain changes such as increased atrophy, which are likely easier to detect than the subtle changes in the CN-to-MCI task. However, for the more difficult task of predicting the change of CN to MCI, the SSL-TOPC model excels, displaying the advantage of spatiotemporal pre-training.

4.3 Future Conversion Prediction

The final and most difficult tasks involve predicting future conversion (within the next 1–2.5 years) from a single MRI scan (Table 1, rows 6 and 7). As mentioned previously, the SSL model with contrastive learning achieves strong results in predicting future conversion from CN to MCI, showing the importance of learning robust spatial features for single-image prediction tasks. All three SSL models perform well for the prediction of MCI to AD. The higher performance can likely be attributed to the temporal pre-training objective, which encourages the model to understand changes over time and therefore strongly contributes to improving prediction of future disease conversion.

5 Conclusion

In this work, we adapt state-of-the-art temporal self-supervised learning approaches for Alzheimer's prediction tasks and include novel extensions to address the challenges of real-world medical images, resulting in three distinct pre-training strategies. We demonstrate that pre-training with increased data and temporal learning objectives can considerably improve downstream performance, with notable improvement when combining temporal order prediction with contrastive learning. Importantly, the SSL-TOPC model trained using both temporal order prediction and contrastive learning achieves superior performance over supervised learning for six of the seven tasks, and is adaptable to fine-tuning across different numbers of input images with varying time intervals, as well as generalizable across numerous downstream tasks. We release all code and pre-trained models publicly, with the intention of providing a generalizable and adaptable model tailored to longitudinal and/or Alzheimer's prediction tasks.

Acknowledgment. This work was supported by the Natural Sciences and Engineering Research Council of Canada, Fonds de Recherche du Quebec: Nature et Technologies, the Canadian Institute for Advanced Research (CIFAR) Artificial Intelligence Chairs program, Calcul Quebec, the Digital Research Alliance of Canada, the Vadasz Scholar McGill Engineering Doctoral Award, Mila - Quebec AI Institute, the International Progressive MS Alliance and the MS Society of Canada.

We also acknowledge all datasets used in this work. HABS-HD: Research reported on this publication was supported by the National Institute on Aging of the National Institutes of Health under Award Numbers R01AG054073, R01AG058533, P41EB015922 and U19AG078109. MCSA: The data contained in this analysis were obtained under research grant from the National Institutes of Health to the Mayo Clinic Study of Aging (U01 AG006786, Ronald Petersen, PI). NIFD: Data collection and sharing for this project was funded by the Frontotemporal Lobar Degeneration Neuroimaging Initiative (National Institutes of Health Grant R01 AG032306). ADNI: Data collection and sharing for the Alzheimer's Disease Neuroimaging Initiative (ADNI) is funded by the National Institute on Aging (National Institutes of Health Grant U19AG024904).

Disclosure of Interests. The authors have no competing interests to declare that are relevant to the content of this article.

References

1. Avants, B., Epstein, C., Grossman, M., Gee, J.: Symmetric diffeomorphic image registration with cross-correlation: evaluating automated labeling of elderly and neurodegenerative brain. Med. Image Anal. **12**(1), 26–41 (2008). https://doi.org/10.1016/j.media.2007.06.004. Special Issue on The Third International Workshop on Biomedical Image Registration – WBIR 2006
2. Billot, B., et al.: Synthseg: segmentation of brain mri scans of any contrast and resolution without retraining. Med. Image Anal. **86**, 102789 (2023). https://doi.org/10.1016/j.media.2023.102789
3. Cardoso, M.J., et al.: Monai: an open-source framework for deep learning in healthcare. arXiv:2211.02701 (2022)
4. Chen, T., Kornblith, S., Norouzi, M., Hinton, G.: A simple framework for contrastive learning of visual representations. In: Proceedings of the 37th International Conference on Machine Learning. ICML'20 (2020)
5. Ericsson, L., Gouk, H., Loy, C.C., Hospedales, T.M.: Self-supervised representation learning: introduction, advances, and challenges. IEEE Signal Process. Mag. **39**(3), 42–62 (2022). https://doi.org/10.1109/MSP.2021.3134634
6. Fathi, S., Ahmadi, M., Dehnad, A.: Early diagnosis of Alzheimer's disease based on deep learning: a systematic review. Comput. Biol. Med. **146**, 105634 (2022). https://doi.org/10.1016/j.compbiomed.2022.105634
7. Frizzell, T.O., et al.: Artificial intelligence in brain MRI analysis of Alzheimer's disease over the past 12 years: a systematic review. Ageing Res. Rev. **77**, 101614 (2022). https://doi.org/10.1016/j.arr.2022.101614
8. Gong, H., Wang, Z., Huang, S., Wang, J.: A simple self-supervised learning framework with patch-based data augmentation in diagnosis of Alzheimer's disease. Biomed. Sig. Process. Control **96**, 106572 (2024). https://doi.org/10.1016/j.bspc.2024.106572

9. Gryshchuk, V., Singh, D., Teipel, S., Dyrba, M.: Contrastive self-supervised learning for neurodegenerative disorder classification. Front. Neuroinform. **19** (2025)
10. He, K., Zhang, X., Ren, S., Sun, J.: Deep residual learning for image recognition. In: 2016 IEEE Conference on Computer Vision and Pattern Recognition (CVPR), pp. 770–778 (2016). https://doi.org/10.1109/CVPR.2016.90
11. Hoopes, A., Mora, J.S., Dalca, A.V., Fischl, B., Hoffmann, M.: Synthstrip: skull-stripping for any brain image. NeuroImage **260**, 119474 (2022). https://doi.org/10.1016/j.neuroimage.2022.119474
12. Jo, T., Nho, K., Saykin, A.J.: Deep learning in Alzheimer's disease: diagnostic classification and prognostic prediction using neuroimaging data. Front. Aging Neurosci. **11** (2019)
13. Kelley, B.J., Petersen, R.C.: Alzheimer's disease and mild cognitive impairment. Neurol. Clin. **25**(3), 577–609 (2007). https://doi.org/10.1016/j.ncl.2007.03.008
14. Misra, I., Zitnick, C.L., Hebert, M.: Shuffle and learn: unsupervised learning using temporal order verification. In: Leibe, B., Matas, J., Sebe, N., Welling, M. (eds.) ECCV 2016. LNCS, vol. 9905, pp. 527–544. Springer, Cham (2016). https://doi.org/10.1007/978-3-319-46448-0_32
15. O'Bryant, S.E., Johnson, L.A., Barber, R.C., Braskie, M.N., Christian, B., Hall, J.R., et al.: The health & aging brain among Latino elders (HABLE) study methods and participant characteristics. Alzheimer's Dementia: Diagn. Assess. Disease Monitor. **13**(1), e12202 (2021)
16. Petersen, R.C., Aisen, P.S., Beckett, L.A., Donohue, M.C., Gamst, A.C., Harvey, D.J., et al.: Alzheimer's disease neuroimaging initiative (ADNI) clinical characterization. Neurology **74**(3), 201–209 (2010)
17. Roberts, R.O., Geda, Y.E., Knopman, D.S., Cha, R.H., Pankratz, V.S., Boeve, B.F., et al.: The mayo clinic study of aging: design and sampling, participation, baseline measures and sample characteristics. Neuroepidemiology **30**(1), 58–69 (2008)
18. Rosen, H.J.: The FTLDNI Study Group: frontotemporal lobar degeneration neuroimaging initiative (FTLDNI). Funded by the National Institute on Aging; coordinated by UCSF and collaborating sites in North America (2010). http://memory.ucsf.edu/research/studies/nifd
19. Shinohara, R.T., et al.: Statistical normalization techniques for magnetic resonance imaging. NeuroImage: Clin. **6**, 9–19 (2014). https://doi.org/10.1016/j.nicl.2014.08.008
20. Thrasher, J., Devkota, A., Tafti, A.P., Bhattarai, B., Gyawali, P.: TE-SSL: time and event-aware self supervised learning for Alzheimer's disease progression analysis. In: Medical Image Computing and Computer Assisted Intervention – MICCAI 2024, pp. 324–333. Springer Nature Switzerland (2024)
21. Tustison, N.J., et al.: N4itk: improved n3 bias correction. IEEE Trans. Med. Imaging **29**(6), 1310–1320 (2010). https://doi.org/10.1109/TMI.2010.2046908
22. Valizadeh, G., Elahi, R., Hasankhani, Z., Rad, H.S., Shalbaf, A.: Deep learning approaches for early prediction of conversion from mci to ad using MRI and clinical data: a systematic review. Arch. Comput. Methods Eng. **32**, 1229–1298 (2025)
23. Xu, D., Xiao, J., Zhao, Z., Shao, J., Xie, D., Zhuang, Y.: Self-supervised spatiotemporal learning via video clip order prediction. In: Computer Vision and Pattern Recognition (CVPR) (2019)
24. Yin, Y., Jin, W., Bai, J., Liu, R., Zhen, H.: Smil-deit:multiple instance learning and self-supervised vision transformer network for early Alzheimer's disease classification. In: 2022 International Joint Conference on Neural Networks (IJCNN), pp. 1–6 (2022). https://doi.org/10.1109/IJCNN55064.2022.9892524

Unstable Prompts, Unreliable Segmentations: A Challenge for Longitudinal Lesion Analysis

Niels Rocholl(✉), Ewoud Smit, Mathias Prokop, and Alessa Hering

Department of Medical Imaging, Radboudumc, Nijmegen, The Netherlands
niels.rocholl@radboudumc.nl

Abstract. Longitudinal lesion analysis is crucial for oncological care, yet automated tools often struggle with temporal consistency. While universal lesion segmentation models have advanced, they are typically designed for single time points. This paper investigates the performance of the ULS23 segmentation model in a longitudinal context. Using a public clinical dataset of baseline and follow-up CT scans, we evaluated the model's ability to segment and track lesions over time. We identified two critical, interconnected failure modes: a sharp degradation in segmentation quality in follow-up cases due to inter-scan registration errors, and a subsequent breakdown of the lesion correspondence process. To systematically probe this vulnerability, we conducted a controlled experiment where we artificially displaced the input volume relative to the true lesion center. Our results demonstrate that the model's performance is highly dependent on its assumption of a centered lesion; segmentation accuracy collapses when the lesion is sufficiently displaced. Our results reveal a fundamental limitation of applying single-timepoint models to longitudinal data. We conclude that robust oncological tracking requires a paradigm shift away from cascading single-purpose tools towards integrated, end-to-end models that are inherently designed for temporal analysis.

Keywords: Longitudinal Analysis · Lesion Segmentation · Computed Tomography · Deep Learning · Registration Error

1 Introduction

Longitudinal imaging is essential for monitoring disease progression and treatment response in oncology. With the number of CT examinations steadily increasing and the global cancer burden projected to rise by 47% by 2040 [1], radiologists face growing pressure to manage high imaging workloads effectively. Cancer patients often undergo multiple CT scans during treatment and follow-up, and response evaluation guidelines such as RECIST 1.1 [2] are frequently applied to measure lesion changes across time. However, lesion measurement and response assessment are still predominantly manual, requiring significant reading time and subject to inter-observer variability. Automatic lesion segmentation models offer

B. Hou and T. S. Mathai (Eds.): LMID 2025, LNCS 16184, pp. 13–23, 2026.
https://doi.org/10.1007/978-3-032-16128-4_2

the potential to reduce this burden by enabling consistent and efficient lesion tracking. Recent studies [4,6,8] have demonstrated that automation of follow-up evaluation can reduce both reading times and disagreement among readers.

While the Universal Lesion Segmentation (ULS [3]) dataset enabled the development of general-purpose lesion segmentation models across a wide range of lesion types, it remains limited to single time points and lacks temporal context. To address this, a new publicly available longitudinal CT dataset [7] was recently introduced, comprising baseline and follow-up scans from melanoma patients undergoing systemic therapy. The dataset includes manual annotations of all visible metastases per patient, offering a valuable benchmark for evaluating lesion-tracking algorithms in longitudinal imaging.

This paper presents an in-depth evaluation of the baseline model from the ULS23 challenge on longitudinal CT scans with annotated metastases [3]. We identify two recurrent and clinically relevant failure modes: (1) spatial drift in lesion localization across timepoints caused by inter-scan registration errors, and (2) the misidentification or complete loss of lesions in follow-up scans as a result. We formalize these issues, quantify their occurrence, and illustrate their impact on lesion tracking. Our findings offer practical insights that can guide the development of more robust and temporally consistent lesion segmentation models.

2 Materials and Methods

To systematically evaluate the ULS23 baseline model in a longitudinal context, we designed a two-part experimental framework. The first experiment assesses the model's overall performance on the publicly available longitudinal CT dataset [7], focusing on both baseline and follow-up scans. The second, more targeted experiment, investigates the influence of registration errors of varying magnitudes on the segmentation output.

2.1 ULS23 Segmentation Model

The ULS23 model [3] is a semi-supervised 3D universal lesion segmentation system, developed within the nnU-Net v1 framework [5]. It employs a 3D Residual Encoder UNet architecture, which was extended from the standard nnU-Net configuration with additional encoder and decoder layers (for a total of seven) and an increased feature channel capacity (upper bound raised from 320 to 384).

A defining characteristic of ULS23 is that it disables the default resampling and input patching strategies typically used in nnU-Net, in order to preserve spatial priors in the input data. Consequently, ULS23 operates on fixed-size Volumes of Interest (VOIs) of $256 \times 256 \times 128$ voxels and, assumes that a single target lesion is located at the center of each VOI. These VOIs are typically generated by cropping around suspected lesion locations.

The training of ULS23 involved a two-stage, semi-supervised process. The model was initially pre-trained using 2D pseudo-masks generated from partially annotated datasets and subsequently fine-tuned on a large, aggregated collection of 3D ground-truth masks. In the second stage, this model was used to generate improved, quality-filtered 3D pseudo-labels. A final fine-tuning step was then performed exclusively on the high-quality 3D ground-truth data, which was derived from VOIs centered on the target lesions. This training strategy, particularly its reliance on accurately centered lesions in VOIs, is a key aspect considered in our analysis of its application to longitudinal data.

2.2 Longitudinal CT Dataset

All 300 scans were acquired at the University Hospital Tübingen (UKT) as part of routine treatment assessment in patients with metastatic melanoma. Each case includes manual 3D segmentations of all malignant lesions, both primary tumors and metastases, at both time points. For each lesion identified on the baseline scan, a coordinate is provided at the center of the lesion (centroid, i.e., the geometric center point). In the corresponding follow-up scan, both the true lesion centroid and a centroid propagated from baseline via conventional image registration are provided (note that propagated centroids are not available for all lesions in the dataset). All images were acquired on CT systems (Siemens Healthineers, Erlangen, Germany) using a standardized whole-body staging protocol during the portal-venous contrast phase, with a reconstructed slice thickness of 3 mm. Crucially, this dataset is publicly available, ensuring our findings are fully reproducible.

The dataset captures a broad range of longitudinal lesion changes, including growth, shrinkage, splitting, merging, complete disappearance, and the appearance of new lesions. Four representative examples from the dataset are shown in Fig. 1, illustrating different lesion progression patterns such as stable, resolved, merged, and progressed lesions.

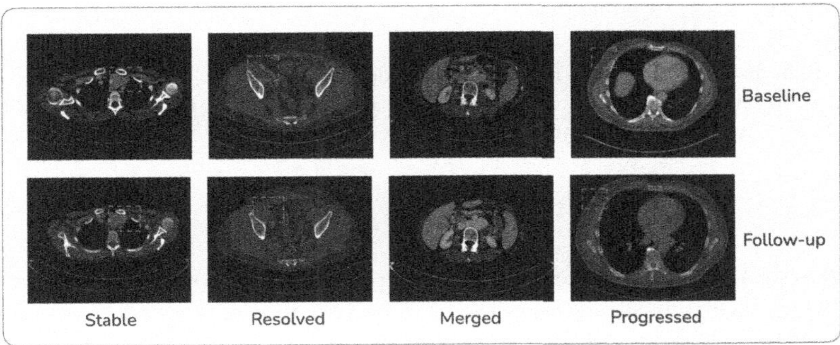

Fig. 1. Four examples from the public Longitudinal CT dataset, showing stable, resolved, merged, and progressed lesions.

2.3 Experimental Setup and Evaluation

We designed two experimental protocols to evaluate the ULS23 model in a longitudinal setting. The first assesses its performance on both baseline and follow-up scans. The second experiment quantifies the model's dependence on its centered-lesion assumption by systematically introducing controlled misalignments and measuring the impact on segmentation performance. The overall workflow of both experiments is visualized in Fig. 2.

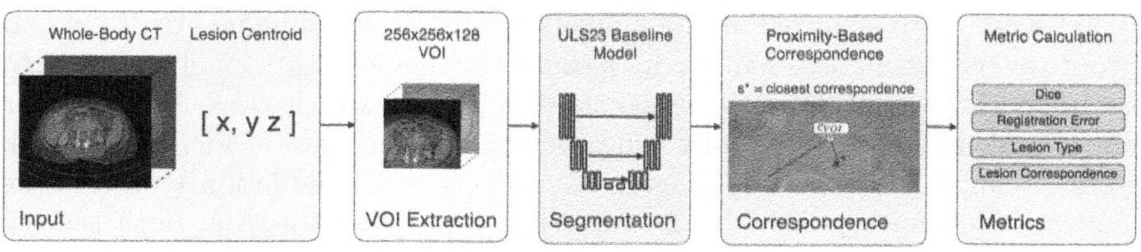

(a) Workflow for the Longitudinal performance evaluation experiment. VOIs are extracted around lesion centroids (propagated for follow-up), segmented by ULS23, correspondence is determined, and metrics are logged.

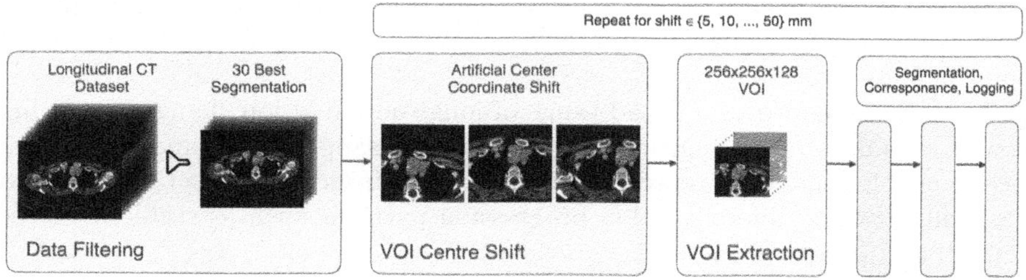

(b) Workflow for the Controlled VOI Displacement Analysis. VOIs from the 30 best-segmented lesions are systematically shifted before segmentation to quantify the impact.

Fig. 2. Schematic overview of the two experimental setups: (a) evaluating ULS23 performance on the full longitudinal dataset, and (b) analyzing ULS23 sensitivity to controlled VOI center displacements.

Longitudinal Performance Evaluation. The first experiment follows a relatively standard four-step segmentation pipeline: VOI extraction, segmentation, lesion correspondence, and metric calculation. This process is schematically outlined in Fig. 2. For a single CT scan, the process unfolds as follows: we iterate through the provided lesion coordinates of the CT volume, each associated with a unique lesion ID. A $256 \times 256 \times 128$ voxel VOI was extracted for each coordinate. For baseline scans, VOIs were centered on the given lesion centroids. For follow-up scans, VOIs were centered on the propagated baseline centroids when available, otherwise, the true follow-up lesion centroid was used as a best-case scenario reference. While centroids for the baseline scans can be assumed to be at the true center of gravity of each lesion, follow-up centroids propagated through registration can be subject to registration error. Inference was then performed on each extracted VOI, yielding a 3D segmentation mask. To establish lesion correspondence, we first identified all distinct connected components within the raw segmentation output. The lesion ID of the current VOI was then assigned to the single connected component whose centroid was closest to the VOI's geometric center. Formally we say that if $S = \{s_1, s_2, \ldots, s_n\}$ is the set of n segmented connected components, $c(s_i)$ is the centroid of component s_i, and c_{VOI} are the coordinates of the VOI center, the assigned component s^* is determined by:

$$s^* = \arg\min_{s_i \in S} \|c(s_i) - c_{\text{VOI}}\|_2$$

As each VOI is processed with the expectation of segmenting a single central lesion, any other segmented components within that VOI were not considered for the metrics associated with that specific lesion.

We defined the outcome of each VOI-level segmentation based on whether the selected component s^* correctly corresponded to the same lesion instance as identified in the baseline scan. This categorization accounts for both successful tracking and the possibility of lesion resolution:

- **Correct Assignment:** The selected component s^* corresponds to the same lesion as identified in the baseline scan. This includes both anatomically and temporally correct matches.
- **True Negative (TN):** No segmentation was produced, and the baseline lesion had resolved, meaning no follow-up lesion was present.
- **Incorrect Assignment:** The component s^* does not correspond to the same lesion as identified in the baseline scan. This includes two common cases: A different lesion was segmented (i.e., wrong match within the VOI), or a segmentation was produced despite the baseline lesion having resolved (i.e., the lesion was no longer present but the model hallucinated one).
- **False Negative (FN):** No component was segmented as s^*, despite the presence of a ground truth target lesion.

For evaluation, we calculated the Dice similarity coefficient for each segmented lesion against its corresponding ground truth. To assess the impact of temporal changes, we compared the distribution of Dice scores between baseline and follow-up scans using the Wilcoxon signed-rank test.

Controlled VOI Displacement Analysis. The second experiment was designed to systematically evaluate the ULS23 model's sensitivity to one of its core assumptions: that lesions are located at the center of the input VOI. The experimental workflow is illustrated in Fig. 2b. Initial analyses of the longitudinal data (Sect. 2.3) indicated that while registration errors often led to off-center lesions, the naturally occurring instances were insufficient for a statistically robust analysis of this specific effect across a controlled range of misalignments. Furthermore, the inherent variability in ULS23's baseline performance on diverse lesions could confound the impact of VOI misalignments.

To create a reliable reference set for the displacement study, we selected the 30 lesions that achieved the highest Dice in the longitudinal-performance experiment. These well-segmented cases provide a consistent, high-quality baseline against which the impact of controlled VOI shifts can be measured more precisely.

For each of these 30 selected lesions, the VOI's designated center coordinate was then artificially shifted from the true lesion centroid. To ensure shifts do not move into zero padded space, they were applied along a direction unique to each lesion, defined by the vector from its true centroid to its CT volume center. These perturbations were applied using magnitudes $\epsilon \in \{0, 5, 10, 15, 20, 25, 30, 35, 40, 45, 50\}$ mm. The $\epsilon = 0$ mm case represents the original, accurately centered VOI for these top-performing lesions. For each resulting shifted coordinate (c_{shifted}), a $256 \times 256 \times 128$ voxel VOI was extracted, centered at c_{shifted}. Subsequently, the ULS23 model performed segmentation, followed by the same lesion correspondence and metric calculation procedures detailed in Sect. 2.3. This allowed for a direct comparison of segmentation quality and correspondence accuracy as a function of controlled VOI displacement from the true lesion center.

3 Results

3.1 Longitudinal Performance Evaluation

Figure 3a highlights the distribution of registration errors for all propagated lesion coordinates. These are calculated by taking the Euclidean distance from the propagated centroid to the centroid of the ground truth mask. Each bar comprises a 1 mm range bin. The histogram shows us that while a substantial portion of propagated centroids has a low registration error, the distribution has a long tail, with a non-trivial number of lesions exhibiting errors greater than 10 mm.

This registration error directly translates to poorer segmentation performance at the follow-up time point. As shown in Fig. 3b, the Dice is significantly lower for the follow-up scans compared to the baseline scans ($p < 0.001$). The distribution of Dice scores for the follow-up scans is also more dispersed, with a larger proportion of very low scores, indicating frequent and severe segmentation failures.

Figure 4a and 4b illustrate the results of our lesion correspondence method. For the baseline cases, we can see that in the majority of cases, the correct lesion is assigned, with some incorrect assignments, as well as some false negatives. In contrast, the follow-up results show a substantial decrease in correct assignments and a notable increase in incorrect assignments, false negatives, and, most notably, true negatives. The number of true negatives (where the model correctly segments nothing) increases from zero at baseline to 31% at follow-up.

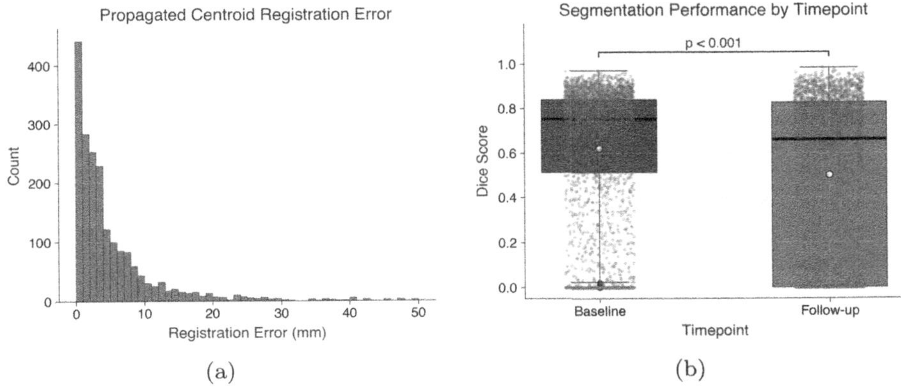

Fig. 3. The impact of registration error on longitudinal segmentation performance. (a) The distribution of registration errors for propagated centroids in follow-up scans reveals a long tail of significant misalignments, with a notable number of lesions displaced by over 10 mm. (b) These errors lead to a substantial drop in Dice score at follow-up compared to baseline ($p < 0.001$), indicating frequent segmentation failures.

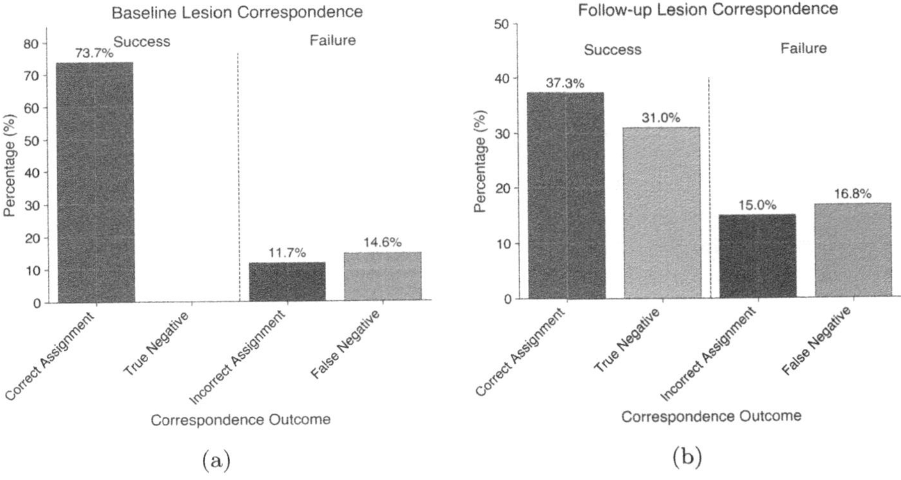

Fig. 4. Lesion correspondence outcomes for (a) baseline and (b) follow-up scans. The rate of clinically critical tracking errors is high at both time points: the combined rate of incorrect assignments and false negatives is 26.3% at baseline and worsens to 31.8% at follow-up. This highlights the fragility of the correspondence method when segmentation is unreliable, a direct consequence of the issues shown in Fig. 3.

3.2 Controlled VOI Displacement Analysis

Figure 5 shows the results of the controlled VOI displacement experiment. Here the true centroid coordinate of the 30 top segmented lesions was systematically shifted, to simulate registration error.

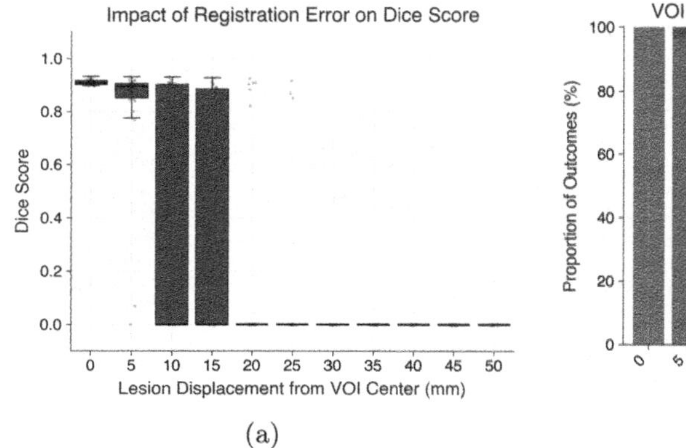

(a)

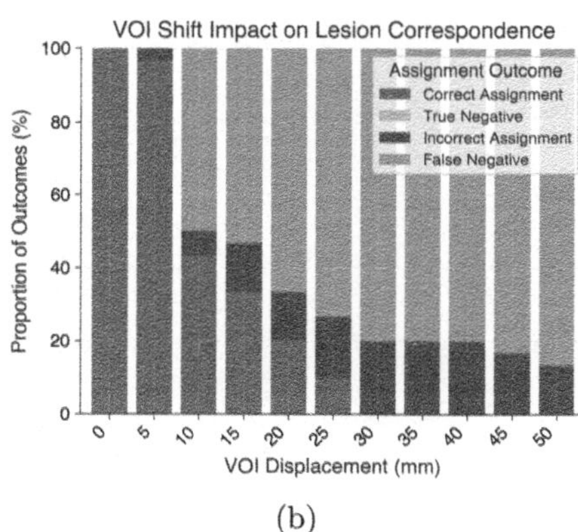

(b)

Fig. 5. Controlled experiment showing ULS23's sensitivity to lesion displacement. (a) The Dice score shows a sharp drop when the lesion is displaced more than 20 mm from the VOI center, indicating complete segmentation failure. (b) Consequently, the rate of correct lesion correspondence falls drastically, with failures dominating even at moderate (10–15 mm) displacements. This controlled experiment confirms that the performance drop seen in real-world follow-up data (Fig. 3) is caused by VOI misalignment.

Figure 5a shows how segmentation performance degrades with increasing (synthetic) registration errors. Instead of a gradual decline, the relationship between the Dice score and the lesion-to-VOI-center distance is defined by a sharp threshold. The model's ability to accurately segment the lesion is high for displacements below approximately 20 mm but deteriorates rapidly beyond that point. For these larger displacements, the Dice score is almost universally near zero, indicating a complete segmentation failure in which the predicted mask shares virtually no mutual information with the ground truth.

Figure 5b illustrates the proportion of lesion correspondence outcomes as a function of artificially introduced displacement. At 0 mm displacement, 100% of assignments are correct. However, the proportion of correct assignments rapidly declines with increasing displacement, falling below 50% by 10 mm. Correspondingly, incorrect assignments and false negatives become more frequent. Beyond a 25 mm shift, there are virtually no correct assignments, with false negatives (i.e., complete failure to segment anything) becoming the dominant outcome.

4 Discussion

The experimental results reveal two fundamental challenges when applying the ULS23 baseline model in a longitudinal context. The first is the model's sensitivity to VOI misalignment, stemming from the violation of its assumption that lesions are always centered in the VOI. The ULS23 model operates under a strong assumption inherited from its training: the lesion is always located at the center of the VOI. Additionally, this bias was reinforced by its training on non-exhaustively annotated data, teaching the model to disregard peripheral lesions.

In a longitudinal setting, this core assumption breaks down. Propagated centroid can suffer from registration errors that frequently displace the target lesion from the VOI's center. This misalignment can result in a significant degradation of segmentation quality, evidenced by the sharp drop in Dice scores from the baseline to the follow-up scans (Fig. 3b). It is important, however, to consider confounding factors beyond geometric misalignment. For instance, changes in lesion morphology and appearance between scans, resulting from treatment or disease progression, are confounding factors that could also contribute to the observed performance degradation. This, combined with the appearance of new and disappearance of old lesions, complicates a direct one-to-one comparison between baseline and follow-up performance.

The controlled experiment detailed in Sect. 2.3 provided a systematic analysis of how VOI-to-lesion displacement impacts segmentation performance. The results, illustrated in Fig. 5a, demonstrate a clear and critical relationship: while segmentation quality (Dice score) is robust to minor shifts, it degrades sharply as the lesion becomes more off-center, often failing completely once a distance threshold is exceeded. This behavior can be naturally explained by the ULS23 model's training assumption to segment the central object, disregarding peripheral structures as background. It is worth noting that some of the larger, artificial displacements may not be realistic for a typical registration algorithm (e.g., shifting a coordinate from one organ to another). However, the results still imply a fundamental principle: the model's success is highly contingent on the accuracy of the guiding coordinate. If this centroid, whether sourced from an automated registration or a manual annotation such as a radiologist's click, is sufficiently far from the actual lesion, the model is designed to ignore it.

The second challenge is the inherent fragility of the center-proximity correspondence method, as its accuracy is entirely dependent on the quality of the segmentation output. The method for establishing lesion correspondence across time points, a center-proximity assignment rule, is conceptually straightforward but operationally fragile. This approach is a cascade of three components: lesion centroid propagation via registration, segmentation of the resulting VOI, and assignment of the lesion ID to the segmented component nearest to the VOI's center. Its effectiveness is therefore critically dependent on the tandem success of accurate registration and correct segmentation. As demonstrated in the previous section, the segmentation step is sensitive to the misalignments introduced by registration errors.

The impact of this dependency is clear when comparing the correspondence outcomes between the baseline (Fig. 4a) and follow-up (Fig. 4b) scans. We observe a significant drop in correct assignments at follow-up. While a substantial fraction of this shift is due to an increase in true negatives, cases where a lesion has resolved and the model correctly segments nothing, the proportion of outright failures also increases. The combined rate of false assignments and false negatives, which represent clinically critical tracking errors, rises from an already high 26.3% at baseline to 31.8% at follow-up. In a clinical setting, where response assessment depends on the consistent tracking of specific lesions, an error rate affecting nearly one-third of follow-up measurements is untenable. Such failures introduce a significant risk of incorrect patient staging and flawed evaluations of treatment efficacy.

The controlled displacement experiment directly illustrates the mechanism behind this failure. As shown in Fig. 5b, as a lesion is artificially shifted further from the VOI center, the rate of incorrect assignments and particularly false negatives increases progressively. This trend is directly related to the segmentation behavior observed in Fig. 5a. Once a lesion is displaced beyond a certain threshold, the model fails to segment it. This results in no connected component being available for the correspondence algorithm, leading inevitably to a false negative. This cascading effect, where registration error leads to segmentation failure, which in turn causes correspondence failure, exposes a fundamental vulnerability in this approach.

5 Conclusion

In this study, we evaluated the ULS23 single time-point segmentation model in a longitudinal context, revealing critical vulnerabilities when applying such isolated analysis tools to serial scans. In follow-up cases, we found the model's performance to be fragile and highly dependent on registration accuracy to satisfy its assumption of a centered lesion. As our experiments demonstrated, moderate misalignments, common in propagated coordinates, frequently lead to complete segmentation failure. This directly undermines our correspondence method, resulting in a high rate of incorrect or missed lesion associations in follow-up scans.

While the model's center bias could be addressed by training with a more varied spatial distribution of lesions, we argue the fundamental issue is the use of cascaded, independent steps which leads to compounding errors. The more robust path forward is to develop unified, end-to-end frameworks that are inherently longitudinal. Recent work has already started to move in this direction, extending nnUNet to integrate the temporal dimension by concatenating multi-timepoint images as additional input channels. [5,9,10] . Future efforts should elaborate on this, and focus on models that jointly optimize for anatomically and temporally consistent lesion correspondence, moving away from chaining single-timepoint tools towards building truly time-aware systems for oncological tracking.

Disclosure of Interests. The authors have no competing interests to declare that are relevant to the content of this article.

References

1. Bray, F., et al.: Global cancer statistics 2022: GLOBOCAN estimates of incidence and mortality worldwide for 36 cancers in 185 countries. CA Cancer J. Clin. **74**(3), 229–263 (2024). https://doi.org/10.3322/caac.21834
2. Eisenhauer, E.A., et al.: New response evaluation criteria in solid tumours: revised RECIST guideline (version 1.1). European Journal of Cancer (Oxford, England: 1990) **45**(2), 228–247 (2009). https://doi.org/10.1016/j.ejca.2008.10.026
3. Grauw, M.J.J.D., et al.: The ULS23 Challenge: a Baseline Model and Benchmark Dataset for 3D Universal Lesion Segmentation in Computed Tomography(2024). https://doi.org/10.48550/arXiv.2406.05231, arXiv:2406.05231 [eess] version: 2
4. Hering, A., et al.: Improving assessment of lesions in longitudinal CT scans: a bi-institutional reader study on an AI-assisted registration and volumetric segmentation workflow. Int. J. Comput. Assist. Radiol. Surg. **19**(9), 1689–1697 (2024). https://doi.org/10.1007/s11548-024-03181-4
5. Isensee, F., Jaeger, P.F., Kohl, S.A.A., Petersen, J., Maier-Hein, K.H.: nnU-Net: a self-configuring method for deep learning-based biomedical image segmentation. Nat. Methods **18**(2), 203–211 (2021). https://doi.org/10.1038/s41592-020-01008-z, publisher: Nature Publishing Group
6. Jacobs, C., et al.: Assisted versus manual interpretation of low-dose CT scans for lung cancer screening: impact on lung-RADS agreement. Radiology. Imaging Cancer **3**(5), e200160 (2021). https://doi.org/10.1148/rycan.2021200160
7. Küstner, T., et al.: Longitudinal-CT (2025). https://fdat.uni-tuebingen.de/records/qwsry-7t837, publisher: University of Tübingen
8. Lu, S.L., et al.: Randomized multi-reader evaluation of automated detection and segmentation of brain tumors in stereotactic radiosurgery with deep neural networks. Neuro Oncol. **23**(9), 1560–1568 (2021). https://doi.org/10.1093/neuonc/noab071
9. Rokuss, M., et al.: LesionLocator: zero-shot universal tumor segmentation and tracking in 3D whole-body imaging (2025). https://doi.org/10.48550/arXiv.2502.20985, arXiv:2502.20985 [cs]
10. Rokuss, M., et al.: Longitudinal segmentation of MS lesions via temporal difference weighting, 10.48550/arXiv.2409.13416. arXiv:2409.13416 [eess] (2024)

Memorisation Bias: AI Predictions for Data Contributors Are Biased Towards Their Health States in the Training Data

Moritz Knolle[1(✉)], Martin J. Menten[1,2,3], Daniel Rueckert[1,2,3], Georgios Kaissis[1,4], and Ben Glocker[3]

[1] Chair for AI in Healthcare and Medicine, Technical University of Munich (TUM) and TUM University Hospital, Munich, Germany
moritz.knolle@tum.de
[2] Munich Center for Machine Learning (MCML), Munich, Germany
[3] Department of Computing, Imperial College London,London, UK
[4] Institute for Machine Learning in Biomedical Imaging, Helmholtz Munich, Neuherberg, Germany

Abstract. AI models are increasingly deployed in clinical practice to assist doctors in diagnostic or screening tasks. However, a critical concern arises from the inherent ability of modern AI models to memorise individual examples from their training datasets. Such memorisation could lead to inaccurate predictions when a model is later used on individuals whose historical data was (potentially unknowingly) used for model training or fine-tuning. In this study, we discover evidence for memorisation bias in two large medical imaging datasets: CheXpert (chest radiography) and Kermany-OCT (optical coherence tomography). Our experiments reveal that a small proportion of data-contributing patients (0.6% and 1.1% for CheXpert/Kermany-OCT, respectively) exhibit significant changes in their predictions on (future) longitudinal evaluation data when their historical data is included for model training. Strikingly, we find that larger, more diagnostically accurate models exhibit increased memorisation bias: for Kermany-OCT, the number of data-contributing patients affected by memorisation increases substantially (from 1.1% to 7.2%) when scaling model size from 1.5 to 80 million parameters. Together, our results raise the question whether the future health outcomes of data-contributing patients could be adversely affected by memorisation bias, i.e., predictions which are biased towards their previous health states.

Keywords: Memorisation · Deep Learning · Medical Imaging

1 Introduction

Given enough training data, artificial intelligence (AI) models can match, or even exceed, the diagnostic accuracy of human experts for standard diagnostic

G. Kaissis and B. Glocker—Shared senior authorship.

B. Hou and T. S. Mathai (Eds.): LMID 2025, LNCS 16184, pp. 24–33, 2026.
https://doi.org/10.1007/978-3-032-16128-4_3

tasks [4,5,7,14,16]. Medical AI thus has immense potential to improve patient outcomes globally by democratising access to high-quality diagnostics [10] and reducing workload for clinical experts.

AI models (i.e., deep neural networks) are highly over-parameterised, statistical models trained to maximise the likelihood of their training data [13]. Specifically, AI models are trained on large-scale datasets to minimise a loss function which penalises incorrect predictions. Despite numerous studies reporting on the remarkable ability of AI models to generalise to unseen data, the process by which AI models learn from examples remains poorly understood. Prior work has demonstrated that AI models are able to generalise to unseen data while often fitting (i.e., memorising) atypical as well as mislabelled training examples [1,21]. Moreover, quantitative evidence suggests that AI models memorise a substantial proportion of their training data [9,22], a phenomenon which might be required for optimal generalisation on long-tailed data distributions [8]. Together, these studies challenge the bias-variance trade-off, the classical understanding of statistical generalisation in AI models.

Medical AI models are typically trained on large collections of anonymised, historical data from data-contributing patients. As such, the fact that an AI model might memorise individual training examples is concerning and raises the question of whether such a model is safe to deploy on data contributors. Take, for example, an AI model trained to detect breast cancer on historical mammography data. This model, which memorised some individual mammograms from the training data as healthy, might provide an incorrect prediction (healthy) when a data-contributing patient returns for cancer screening with a mammogram that now shows a small malignant tumour. This poses a fundamental dilemma, as the question arises whether data contributors would need to be excluded from AI being used to assess their future data. It is an ethical dilemma, as data contributors would not benefit from advanced diagnostic tools, and a very practical one, as we may not know who contributed training data due to the fact that datasets are typically anonymised to protect patient privacy.

With this work, we present the first investigation of neural network memorisation on longitudinal medical data. Concretely, we investigate *memorisation bias*, i.e., whether the inclusion of an individual patient's historical data in an AI model's training dataset significantly influences the predictions on their future data from follow-up visits. We find significant evidence for memorisation bias on two large medical image datasets of chest radiographs and optical coherence tomography images. Our experiments were inspired by state-of-the-art approaches from the privacy auditing community, specifically membership inference attacks [2,17,20]. Crucially, however, our results go beyond privacy concerns for sharing personal medical data to train AI models. We demonstrate that real-world harm could come to data contributors upon model deployment in the form of predictions that are biased towards their health states in the training data.

2 Methods

We propose a simple approach to measure memorisation on longitudinal follow-up data which is described in detail below.

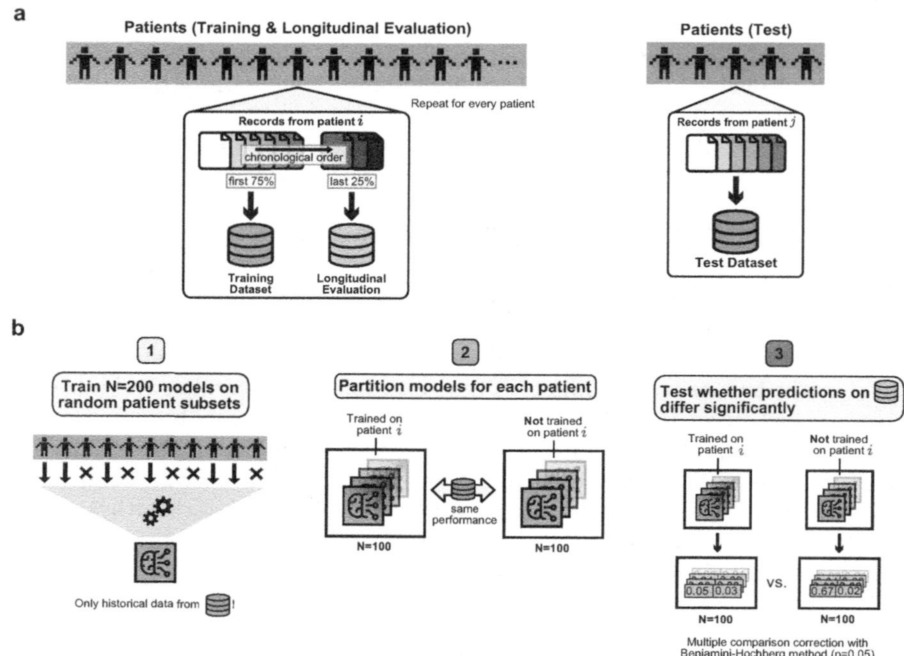

Fig. 1. Schematic of our dataset split strategy (a) and experimental setting (b). **a**, custom three-way dataset split into: training dataset, longitudinal evaluation dataset and test dataset. Longitudinal evaluation split serves to simulate a deployment setting where data-contributing patients encounter a model that was trained on their historical data. Generalisation performance is measured on a conventional test set comprising unseen data from unseen patients. **b**, our experimental setup for measuring memorisation on longitudinal follow-up data from data contributors involves three steps: (1) training $N = 200$ disease detection models on randomly selected subsets of patients (historical training data only), (2) partitioning models into those trained and those not trained on the respective patient and (3) testing whether the predictions on a patient's longitudinal evaluation data differ significantly between these two groups of models using a non-parametric multivariate hypothesis test.

2.1 Datasets

We perform experiments on two large-scale medical imaging datasets:

CheXpert [11]. A large chest-radiograph dataset from Stanford Hospital comprising 224,316 chest radiographs from 65,420 patients. CheXpert comes with structured labels indicating the presence of 14 common thoracic conditions; multiple conditions can be present in a given image. Models for this dataset were trained with data augmentation (random horizontal flipping, random pixel shifts, and random rotations) to detect the presence of all 14 conditions (14-class multi-label classification). We only use frontal radiographs and exclude any radiographs from other views from the training and test datasets. On the official test dataset, a WRN-28-2 [19] trained on the historical training data from a random 50% subset of the data contributing patients achieves a macro-average AUC score of 0.876 across the CheXpert competition classes: Cardiomegaly, Edema, Consolidation, Atelectasis and Pleural Effusion.
OCT-Kermany [12]. A large dataset of retinal optical coherence tomography (OCT) images (foveal cut of three-dimensional volume) comprising 83,484 images from 4,590 patients. Images come with structured labels indicating the progression status of age-related macular degradation (AMD): healthy, drusen, diabetic macular edema and choroidal neovascularisation. Models for this dataset were trained with data augmentation (random contrast/brightness changes and random rotations) to categorise images by AMD stage (4-class multi-class classification). On the official test dataset, a WRN-28-2 trained on the historical training data from a random 50% subset of the data-contributing patients achieves a macro-average AUC score of 0.991 across all classes.

2.2 Dataset Split Strategy and Split Details

We perform a custom three-way data split for both investigated datasets (see Fig. 1 a). Specifically, for each patient in the training dataset, we first sort all images in chronological order. Then, given that a patient has more than four images, we use the first 75% as historical training data and the last 25% of images for longitudinal evaluation. If a patient has fewer than four but more than one image(s), we use the last image for longitudinal evaluation. Thus, every image in the longitudinal evaluation dataset belongs to a patient who has at least one image from a previous encounter in the training dataset. Note that the converse is not generally true since there are patients who contribute only a single image and are thus excluded from longitudinal evaluation.

We measure generalisation performance on the official test sets (conventional patient-wise splits); the test sets thus comprise unseen data from unseen patients. See Table 1 below for detailed information on the dataset splits resulting from our split strategy for both of the investigated datasets.

2.3 Model Training Details

Models were trained using state-of-the-art training techniques. Concretely, models were trained using stochastic gradient descent (SGD) with momentum, exponential moving parameter average (EMA), weight decay, data augmentation, and a learning rate schedule in the form of cosine decay with linear warm-up.

Table 1. Dataset split information.

Dataset	Training (historical)		Longitudinal Evaluation		Test	
	#Patients	#Images	#Patients	#Images	#Patients	#Images
CheXpert	64, 200	139, 185	30, 829	50, 957	500	519
Kermany-OCT	4, 590	68, 972	4, 281	14, 512	633	1, 000

Hyperparameter values were determined through a random search to maximise diagnostic performance (macro average AUC) on unseen test data (see Table 2 for resulting values). Images were resized to 64×64 and rescaled to have values within $[-1, 1]$.

Table 2. Model training hyperparameters. α, λ, μ and γ denote learning rate, weight decay, momentum, and EMA decay rate, respectively. Training time was measured on a single A100 GPU.

Dataset	α	λ	μ	γ	Batch Size	Epochs	Time (min)
CheXpert	1.0	10^{-4}	0.9	0.9995	1024	60	24
Kermany-OCT	5×10^{-3}	10^{-4}	0.9	0.99	256	100	3

2.4 Measuring Memorisation

The proposed approach for measuring memorisation comprises three steps (depicted in Fig. 1 b).

Step 1: Training on Random Patient Subsets. We train $N = 200$ models on the historical data from randomly selected 50% patient subsets. Thus, the training dataset for each of these models either contains or does not contain all historical data of a given data-contributing patient. To enable partitioning the models, for each data contributor, by the inclusion/exclusion of their historical data, we log the patient identifiers of the selected patients for each training run.

Step 2: Partitioning Trained Models by Patient. After training is completed, we simply partition the models into:

- (**IN** models): models which saw the data contributor's historical data during training
- (**OUT** models): models which did **NOT** see the data contributor's historical data during training (OUT models)

For each data contributor, this yields $N = 100$ models per group. Diagnostic performance differences (as measured on the official test sets) between the models are negligible. Note that, since the models are trained on random patient subsets, the effect of including/excluding other patients' data is marginalised out and is thus negligible.

Step 3: Statistical Significance Testing. Next, we compare the predicted probabilities of the IN/OUT models on the longitudinal evaluation data. Concretely, we perform two-sample multivariate hypothesis tests [18] on the predicted probabilities for each image in the longitudinal evaluation dataset using the following hypotheses:

- $\mathbf{H_0}$: $\Pr_{\text{IN}}(x) = \Pr_{\text{OUT}}(x)$,
- $\mathbf{H_1}$: $\Pr_{\text{IN}}(x) \neq \Pr_{\text{OUT}}(x)$,

where $\Pr_{\text{IN/OUT}}(x)$ refers to predicted probabilities (for all classes) on image x provided by IN/OUT models. We perform multiple comparison correction across test results for all images in a respective longitudinal evaluation set using the Benjamini-Hochberg method with a false discovery rate of $p = 0.05$.

3 Results

3.1 Evidence for Memorisation Bias

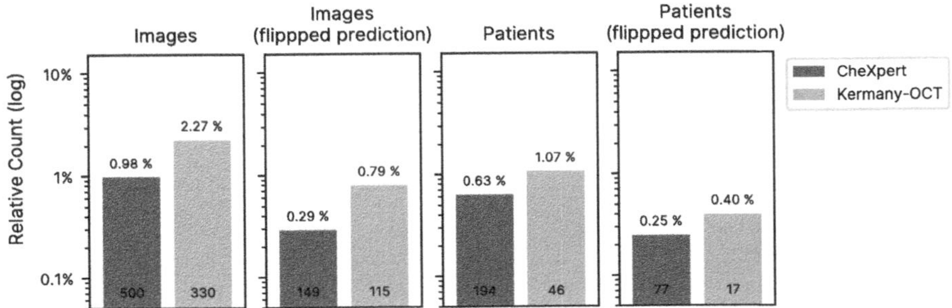

Fig. 2. Significant differences in predictions on longitudinal evaluation data. From left to right, sub-panels show the relative number of: (**a**) images where predicted probabilities differ significantly, (**b**) images where predicted probabilities differ significantly and the prediction flips, (**c**) patients with at least one image where a significant difference in occurs, (**d**) patients with at least one image where a significant difference in prediction occurs and the prediction flips. Absolute counts are shown at the bottom of each bar. Multiple comparison correction was performed using the Benjamini-Hochberg method with a false discovery rate of $p = 0.05$.

We found $N_C = 500\,(0.98\%)$ and $N_K = 330\,(2.27\%)$ images (CheXpert and Kermany-OCT, respectively) from the longitudinal evaluation datasets where the predicted probability by IN and OUT models differs significantly (see Fig. 2). These significant changes in predicted probability led to a flipped prediction (i.e., a change in the predicted class) for $N_C = 149\,(0.29\%)$ and $N_K = 115\,(0.79\%)$ images. The significant changes in predicted probability affected $N_C = 194\,(0.63\%)$ and $N_K = 46\,(1.07\%)$ patients (CheXpert and Kermany-OCT respectively) of which $N_C = 77(0.25\%)$ and $N_K = 17\,(0.40\%)$ experienced a flipped prediction on at least one of their longitudinal evaluation images.

As a sanity check, we also compared the predictions of two distinct sets of OUT models in the same way as described above. This yielded zero longitudinal evaluation images with significantly different predictions for both investigated datasets.

3.2 Memorisation Bias Increases with Model Size

Next, we investigate the impact of increased model capacity on memorisation bias for the Kermany-OCT dataset. Concretely, we trained $N = 200$ models of the following architectures: WRN-28-2, WRN-28-5 and ViT-B/16 [6] (1.5, 10 and 80 million parameters, respectively). Increasing model size led to a sustained increase in diagnostic performance, i.e., a monotonic increase in macro AUROC score on the test set (mean±SD): 0.991±0.0215, 0.996±0.0076 and 0.999±0.0004, respectively.

Alarmingly, we found that larger, more diagnostically accurate models exhibit substantially increased memorisation bias compared to smaller models (see Fig. 3). Most notably, we find that the number of data-contributing patients affected by a significant change in predicted probability on their follow-up data increases nearly 7-fold from 1.07% to 7.19% for WRN-28-2 and ViT-B/16 models. Similarly, the number of data-contributing patients with a significant change that led to a flipped prediction increased from 0.40% to 1.75% (more than a 4-fold increase).

4 Discussion

With this study, we present initial evidence suggesting that medical AI models provide predictions for data contributors that are biased towards historical health states in the training data. Concretely, we found that a small proportion of data-contributing patients ($\sim 1\%$) are affected by such memorisation; these patients exhibit significant changes in predicted probability on their longitudinal evaluation images when their historical data is included for model training. A subset of these patients (0.25 to 0.4%) is impacted directly by a change in the predicted class (disease) for at least one of their longitudinal evaluation images. Together, these results raise concerns that the deployment of AI models could lead to adverse health outcomes for data contributors. Medical AI models are

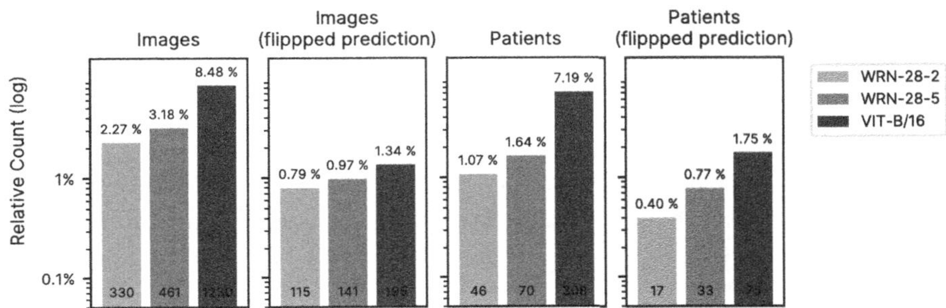

Fig. 3. Larger, more diagnostically accurate models exhibit increased memorisation bias. Statistical significance testing results for $N = 200$ models of increasing size: WRN-28-2, WRN-28-5 and ViT-B/16 (1.5, 10 and 80 million parameters, respectively). Larger models show increased diagnostic performance on the test set (mean$\pm$SD): 0.991 ± 0.0215, 0.996 ± 0.0076 and 0.999 ± 0.0004, respectively. Panels show relative counts as explained in the caption of Fig. 2; absolute counts are shown at the bottom of each bar. We corrected for multiple comparisons using the Benjamini-Hochberg method with a false discovery rate of $p = 0.05$.

often deployed in triage or screening settings. Thus, even small changes in the predicted risk for a disease could lead to adverse health outcomes for data contributors due to, e.g., delayed treatment.

Alarmingly, the findings from our model scaling experiments indicate that larger, more diagnostically accurate models exhibit much more memorisation bias than smaller models. This finding is in line with results from related work on memorisation and privacy auditing [2,3,15,20,22]. Unfortunately, recent theoretical results [8] suggest that this trend of increased memorisation from larger, more performant AI models could be inevitable. The very process by which AI models generalise optimally on long-tailed data distributions (many naturally-occurring data distributions are long-tailed, e.g., medical images, text, or video) requires a model to memorise atypical examples from rare sub-populations [8,9,22]. This indicates that patients from minority groups or those with otherwise atypical data could be disproportionately impacted by memorisation bias. We will investigate this question in future work.

Our work does not come without limitations. Primarily, we were limited by the fact that CheXpert and Kermany-OCT do not provide information on image acquisition time. It was thus not possible for us to assess the time difference between memorised historical training data and the longitudinal evaluation images. Furthermore, our results are currently restricted to only two datasets. In future work, we are planning to investigate memorisation bias on other datasets which contain the necessary image acquisition time information.

5 Conclusion

We present evidence suggesting that the inclusion of a patient's historical data into an AI model training dataset can significantly impact the predictions on data from future follow-up visits. Our findings raise concerns that data contributors could receive biased, inaccurate predictions, which could lead to adverse health outcomes when an AI model is deployed on a population that contributed to its training data.

Acknowledgments. This work was partially funded by the Konrad Zuse School of Excellence in Reliable AI (relAI). M.J.M. is funded by the German Research Foundation under project 532139938. B.G. received support from the Royal Academy of Engineering as part of his Kheiron/RAEng Research Chair.

Disclosure of Interests. B.G. is part-time employee of DeepHealth. No other competing interests.

References

1. Arpit, D., et al.: A closer look at memorization in deep networks. In: International Conference on Machine Learning, pp. 233–242. PMLR (2017)
2. Carlini, N., Jagielski, M., Zhang, C., Papernot, N., Terzis, A., Tramer, F.: The privacy onion effect: memorization is relative. Adv. Neural. Inf. Process. Syst. **35**, 13263–13276 (2022)
3. Carlini, N., et al.: Extracting training data from diffusion models. In: 32nd USENIX Security Symposium (USENIX Security 23), pp. 5253–5270 (2023)
4. De Fauw, J., et al.: Clinically applicable deep learning for diagnosis and referral in retinal disease. Nat. Med. **24**(9), 1342–1350 (2018a)
5. De Fauw, J., et al.: Clinically applicable deep learning for diagnosis and referral in retinal disease. Nat. Med. **24**(9), 1342–1350 (2018b)
6. Dosovitskiy, A., et al.: An image is worth 16x16 words: transformers for image recognition at scale. In: International Conference on Learning Representations (2021)
7. Esteva, A., et al.: Dermatologist-level classification of skin cancer with deep neural networks. Nature **542**(7639), 115–118 (2017)
8. Feldman, V.: Does learning require memorization? a short tale about a long tail. In: Proceedings of the 52nd Annual ACM SIGACT Symposium on Theory of Computing, pp. 954–959 (2020)
9. Feldman, V., Zhang, C.: What neural networks memorize and why: discovering the long tail via influence estimation. Adv. Neural. Inf. Process. Syst. **33**, 2881–2891 (2020)
10. Fleming, K.A., et al.: The lancet commission on diagnostics: transforming access to diagnostics. Lancet **398**(10315), 1997–2050 (2021)
11. Irvin, J., et al.: Chexpert: a large chest radiograph dataset with uncertainty labels and expert comparison. In: Proceedings of the AAAI Conference on Artificial Intelligence, vol. 33, pp. 590–597 (2019)
12. Kermany, D.S., et al.: Identifying medical diagnoses and treatable diseases by image-based deep learning. Cell **172**(5), 1122–1131 (2018)

13. LeCun, Y., Bengio, Y., Hinton, G.: Deep learning. Nature **521**(7553), 436–444 (2015)
14. McKinney, S.M.: International evaluation of an AI system for breast cancer screening. Nature **577**(7788), 89–94 (2020)
15. Nasr, M., et al.: Scalable extraction of training data from (production) language models. CoRR abs/2311.17035 (2023). https://doi.org/10.48550/arXiv.2311.17035
16. Ouyang, D., et al.: Video-based AI for beat-to-beat assessment of cardiac function. Nature **580**(7802), 252–256 (2020)
17. Shokri, R., Stronati, M., Song, C., Shmatikov, V.: Membership inference attacks against machine learning models. In: 2017 IEEE Symposium on Security and Privacy (SP), pp. 3–18. IEEE (2017)
18. Székely, G.J., Rizzo, M.L.: Energy statistics: a class of statistics based on distances. J. Stat. Plann. Inference **143**(8), 1249–1272 (2013)
19. Zagoruyko, S., Komodakis, N.: Wide residual networks. arXiv preprint arXiv:1605.07146 (2016)
20. Zarifzadeh, S., Liu, P., Shokri, R.: Low-cost high-power membership inference attacks. In: Salakhutdinov, R., et al. (eds.) Proceedings of the 41st International Conference on Machine Learning. Proceedings of Machine Learning Research, vol. 235, pp. 58244–58282. PMLR (2024). https://proceedings.mlr.press/v235/zarifzadeh24a.html
21. Zhang, C., Bengio, S., Hardt, M., Recht, B., Vinyals, O.: Understanding deep learning (still) requires rethinking generalization. Commun. ACM **64**(3), 107–115 (2021)
22. Zhang, C., Ippolito, D., Lee, K., Jagielski, M., Tramèr, F., Carlini, N.: Counterfactual memorization in neural language models. In: Thirty-seventh Conference on Neural Information Processing Systems (2023). https://openreview.net/forum?id=67o9UQgTD0

Semi-disentangled Spatiotemporal Implicit Neural Representations of Longitudinal Neuroimaging Data for Trajectory Classification

Agampreet Aulakh[1(✉)], Nils D. Forkert[2], and Matthias Wilms[3]

[1] Biomedical Engineering Graduate Program, University of Calgary, Calgary, Canada
agampreet.aulakh@ucalgarv.ca

[2] Department of Radiology, University of Calgary, Calgary, Canada

[3] Department of Radiology, University of Michigan, Ann Arbor, USA

Abstract. The human brain undergoes dynamic, potentially pathology-driven, structural changes throughout a lifespan. Longitudinal Magnetic Resonance Imaging (MRI) and other neuroimaging data are valuable for characterizing trajectories of change associated with typical and atypical aging. However, the analysis of such data is highly challenging given their discrete nature with different spatial and temporal image sampling patterns within individuals and across populations. This leads to computational problems for most traditional deep learning methods that cannot represent the underlying continuous biological process. To address these limitations, we present a new, fully data-driven method for representing aging trajectories across the entire brain by modelling subject-specific longitudinal T1-weighted MRI data as continuous functions using Implicit Neural Representations (INRs). Therefore, we introduce a novel INR architecture capable of partially disentangling spatial and temporal trajectory parameters and design an efficient framework that directly operates on the INRs' parameter space to classify brain aging trajectories. To evaluate our method in a controlled data environment, we develop a biologically grounded trajectory simulation and generate T1-weighted 3D MRI data for 450 healthy and dementia-like subjects at regularly and irregularly sampled timepoints. In the more realistic irregular sampling experiment, our INR-based method achieves 81.3% accuracy for the brain aging trajectory classification task, outperforming a standard deep learning baseline model (73.7%).

Keywords: Longitudinal Neuroimaging · Implicit Neural Representation · Brain Aging

1 Introduction

Subject-specific longitudinal neuroimaging data, such as repeated T1-weighted Magnetic Resonance Imaging (MRI), are vital for data-driven studies of healthy

B. Hou and T. S. Mathai (Eds.): LMID 2025, LNCS 16184, pp. 34–45, 2026.
https://doi.org/10.1007/978-3-032-16128-4_4

brain aging and pathological deviations associated with neurodegenerative diseases. These data enable the characterization of subject- and population-specific trajectories of typical and atypical brain changes over time. For example, in the case of Alzheimer's Disease (AD), longitudinal data have been instrumental in identifying the morphological signatures of different disease subtypes [19] and can also be used to detect conversion from normal aging to dementia early on [2].

Despite their value and increasing availability through long-term studies, such as the Alzheimer's Disease Neuroimaging Initiative (ADNI) cohorts [14], acquiring longitudinal images in a fully uniform manner across larger cohorts is extremely difficult. In practice, the number of images acquired for each participant usually differs, the data are typically sampled at inconsistent timepoints, and if acquisition is conducted over decades, it is reasonable to expect considerable differences in spatial resolution and image quality over time [23]. All of these aspects make the efficient and meaningful processing of information available at the discretized image level highly challenging.

To circumvent some of these issues, a substantial body of research has been dedicated to extracting scalar image-derived biomarkers, such as cortical thickness or ventricular volume, to analyze brain aging trajectories through mixed-effects models (*e.g.*, [13]). While such methods produce clinically relevant results and are computationally efficient, they lack the flexibility needed to capture the full range of possible subject-specific trajectories and heavily rely on arbitrary design decisions, such as the parcellation scheme used for biomarker extraction. Moreover, these methods only give rise to coarse, biomarker-level trajectories and do not capture important interactions between image features.

Beyond traditional biomarker analyses, several deep learning methods have been developed to model brain aging trajectories directly using imaging data. For instance, it has been shown that trajectories can be modelled as embeddings between image pairs within a latent space that is organized by subjects with similar dynamics [15,16]. Other methods [2,17] first use neural networks to extract features from each image within a longitudinal dataset, and then classify each timepoint as healthy or dementia-like. Critically, these approaches are not able to directly parameterize the full longitudinal aging trajectory [15,16], or rely on intermediate, imaging feature-based representations for analyses [2,17].

We aim to address the challenges associated with processing variable, subject-specific, longitudinal neuroimaging data and the shortcomings of current deep learning-based analysis methods in a unified way to make the best use of the full discrete image information available. Specifically, we propose to directly parameterize individual brain aging trajectories by modelling longitudinal imaging data in a space- and time-continuous manner via Implicit Neural Representations (INRs). These INRs are simple neural networks that learn to approximate a continuous function mapping from the domain of coordinates to their associated signal value. Such data representations have been previously adapted to and used for the classification of 3D object images/point clouds [3,12], high-resolution natural images [1,9], and, more recently in a medical context, low-resolution

2D scans [7]. Lately, INRs have also been popularized for medical image registration purposes [22,25]. Applied to longitudinal neuroimaging, the space- and time-continuous nature of INRs overcomes the problems associated with different temporal (and spatial) sampling patterns and enables a fully data-driven, learning-based representation of longitudinal data. INRs can be flexibly adapted to subject-specific data as the parameter count of these networks is independent of the number of subjects studied and amount of data available per subject. Furthermore, we show for the first time that their parameters can be used for the classification of brain aging trajectories (normal aging *vs.* AD-like aging).

Thus, the main contributions of this paper are threefold: First, we present a new INR architecture with semi-disentangled spatial and temporal parameters to model subject-level aging trajectories across the entire brain from longitudinal T1-weighted MRI data. We show that these INRs can be constructed using data with irregular temporal sampling within subjects and across a cohort. Second, we propose an efficient method for directly classifying the trajectories encoded by these representations. Third, to perform a thorough proof-of-concept evaluation of our methods in a fully controlled manner, we introduce a synthetic brain aging trajectory simulation framework capable of generating longitudinal 3D imaging data of realistic healthy and AD-like aging patterns.

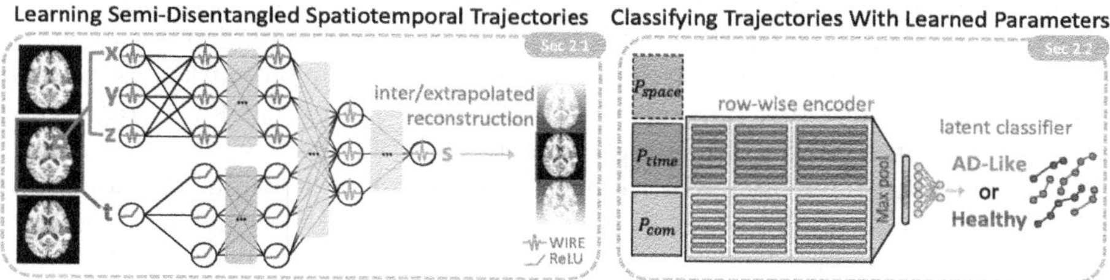

Fig. 1. Overview of proposed methods. Left: Subject-specific trajectory modelling with a semi-disentangled spatiotemporal INR. Right: Organization of INR parameters in space-only, time-only, and combined streams, their row-wise embedding in latent space, and subsequent trajectory classification.

2 Materials and Methods

We assume the biological process of brain aging follows a continuous, subject-specific trajectory $A(\mathbf{c}, t)$ and consider individual T1-weighted MRI scans (longitudinal data) as discrete samples of the structural changes over time t at each spatial coordinate $\mathbf{c}$. Our goal is then twofold: (1) To continuously approximate the underlying trajectory in space and time through an INR trained on discrete image samples. (2) To build a computationally efficient trajectory classifier that operates on the continuous INRs.

Formally, we consider data from M subjects such that each subject $m \in \{1, \ldots, M\}$ has a temporal sequence of N_m scans denoted as $\mathcal{D} = \{(\mathbf{I}_i, t_i)\}_{i=1}^{N_m}$. Here, $\mathbf{I}_i \in \mathbb{R}^{X \times Y \times Z}$ is the 3D image tensor acquired at a discrete sampling time $t_i \in \mathcal{T} \subset \mathbb{R}$. Each image corresponds to a discrete sampling pattern of spatial coordinates $\mathbf{c} = (x, y, z)^T \in \mathcal{C} \subset \mathbb{R}^3$, with voxels representing T1-weighted MRI signal intensity values $s \in \mathcal{S} \subset \mathbb{R}$. While this data $\mathcal{D}$ is inherently discrete, we derive a continuous approximation of the brain aging trajectory $f_\theta(\mathbf{c}, t) \approx A(\mathbf{c}, t)$ through an INR $f_\theta(\cdot)$ with parameters θ that fully parameterize the underlying continuous process (see Sect. 2.1). Assuming a fixed INR architecture across all M subjects, we then propose a trajectory classifier that uses the subject-specific parameters θ_m as input (see Sect. 2.2). The framework is illustrated in Fig. 1.

2.1 Semi-disentangled Spatiotemporal Implicit Neural Representations

Following common practice [21], we use a multilayer perceptron (MLP) to instantiate an INR. Specifically, we construct a subject-specific INR $f_\theta : \mathcal{C} \times \mathcal{T} \to \mathcal{S}$ mapping the input image domain tuples $(\mathbf{c}, t)$ to image signal values $s \in \mathcal{S}$. For an L-layer INR with hidden size H, we use residual connections such that the output of each layer l is defined as:

$$\mathbf{y}_l = \sigma\left(\mathbf{W}_l \mathbf{y}_{l-1} + \mathbf{b}_l\right) + \mathbf{y}_{l-1}, \tag{1}$$

where σ is an activation function, and $\mathbf{W}_l \in \mathbb{R}^{H \times H}$ and $\mathbf{b}_l \in \mathbb{R}^{H \times 1}$ are the weights and biases. These parameters are optimized for a subject m by minimizing an MSE loss $\mathcal{L} = \frac{1}{J \cdot N_m} \sum_{i=1}^{N_m} \sum_{j=1}^{J} (f_\theta(\mathbf{c}_j, t_i) - \mathbf{I}_i[\mathbf{c}_j])^2$ over a randomly selected subset of J voxels for each $\mathbf{I}_i$ in $\mathcal{D}$.

While fitting an INR to a longitudinal dataset $\mathcal{D}$ is generally challenging due to its usually high spatial and low temporal resolution, we make two key observations to simplify this optimization task: (1) Neuroimaging data contains high spatial frequencies as regions of the brain have unique structures (*e.g.*, cortical folding). (2) The frequencies of the temporal domain remain low as one can expect gradual, aging or pathology-driven structural change over time.

We, therefore, propose to disentangle the spatial and temporal processing early on by separating layers receiving $\mathbf{c}$ as input from those receiving t, and to only combine spatial and temporal information at the end (see Fig. 1). This results in a novel, semi-disentangled INR architecture where each stream can be independently biased to capture a different range of frequencies using activation functions σ. Namely, we use wavelet activations (*i.e.*, WIRE [21]) for the spatial and combined streams as they are the state-of-the-art for capturing high-frequency signals, and use ReLU activations for the temporal stream. The spectral biases of both activations have been studied in the past, with multiple works providing results of ablation experiments that complement our design [5,20,21]. With this architecture, we partially decouple space-only, time-only, and space-time combined parameters, which enables us to selectively ignore or emphasize each stream in downstream analysis (*e.g.*, trajectory classification).

This is advantageous for limiting morphology bias (*e.g.*, exploiting shortcuts in space-only parameters) when studying biological processes that are largely time-dependent, such as $A(\mathbf{c}, t)$.

2.2 Brain Aging Trajectory Classification

It is challenging to perform downstream analyses directly on INRs as they may comprise a large number of parameters θ_m (on the order of millions). Then, even simple operations such as applying an MLP to flattened parameters can lead to prohibitive memory requirements and result in poor generalization due to permutation and scale symmetries in parameter space [11].

Therefore, we propose to efficiently solve this brain aging trajectory classification challenge in two steps. First, we aim to reduce the effects of parameter symmetries by using a data-driven initialization strategy where we optimize an INR $f_{\theta^*}(\cdot)$ across all subjects. The resulting initial parameters θ^* are optimized for reconstructing an average dataset across the temporal range and can then be quickly adapted to form a subject-specific representation θ_m. Second, we implement a strategy to organize parameters θ_m for efficient processing and use an encoder to map them to a latent space optimized for classification. For each hidden layer of the INR network, we stack the weights and biases, $\mathbf{W}_l$ and $\mathbf{b}_l^T$, to produce layer-specific transformation parameters $\mathbf{P}_l \in \mathbb{R}^{(H+1)\times H}$. These are then grouped by their respective processing streams to form spatial parameters $\mathbf{P}_{\text{space}} \in \mathbb{R}^{L_S(H+1)\times H}$, temporal parameters $\mathbf{P}_{\text{time}} \in \mathbb{R}^{L_T(H+1)\times H}$, and space-time combined parameters $\mathbf{P}_{\text{com}} \in \mathbb{R}^{L_C(H+1)\times H}$, for L_S, L_T, and L_C number of layers in each stream. To build an efficient classifier, we employ an encoder inspired by [12] consisting of three blocks of linear layers with batch norm and ReLU activation (see Fig. 1). This encoder processes the input parameters matrix row-wise, akin to processing each neuron's weight vector or each layer's bias vector independently, while sequentially expanding the dimension H to H'. Therefore, the same encoder parameters are learned and used to process each row of INR parameters. A latent representation $\mathbf{l} \in \mathbb{R}^{H'}$ is obtained by max pooling across each column, and is then passed on through fully connected layers, with ReLU activation and dropout, which are trained for trajectory classification using a binary cross-entropy loss. Ultimately, this process enables us to build a classifier that is capable of processing any combination of INR parameter streams.

2.3 Brain Aging Trajectory Simulation

For an initial proof-of-concept evaluation, we design a new framework for simulating brain aging trajectories to remove the effects of uncontrollable imperfections (*e.g.*, registration errors and misalignment, movement artifacts, scanner-related biases) associated with real longitudinal imaging data. Specifically, we develop a method for artificially generating a sequence of longitudinal 3D MRI data with accelerated, AD-like brain aging by systematically altering data from simulated subjects with healthy brain aging. This leads to perfect counterfactual scenarios where a simulated subject with the same base morphology can be rendered with either healthy or pathological aging, thus allowing us to identify morphology-driven classification bias.

Therefore, we first simulate subject-specific trajectories by sampling longitudinal MRI data from a recent 3D (age-)conditional diffusion model [24] trained on cross-sectional data from more than 5,000 cognitively healthy subjects in the UK Biobank. This model has shown a competitive performance for biological brain age prediction and excels at generating high-resolution, realistic 3D T1-weighted MR images for given chronological age values. With this diffusion model, we are able to generate diverse longitudinal data for artificial subjects between 50 and 90 years of age with any temporal sampling scheme.

Given its training data, the diffusion model only generates healthy aging trajectories. To simulate a trajectory classification task (healthy *vs.* AD-like accelerated aging), we systematically alter the sampled healthy trajectories by introducing a non-linear deviation mapping between chronological age and biological brain age. In the diffusion model-sampled data, both values are equal and we mimic the brain age drift commonly observed in AD [6] by reassigning biologically older brains to lower chronological ages. Driving this deviation in a realistic manner, we model the change in biological brain age, $\frac{dBA}{dt}$, in relation to chronological age t with an ordinary differential equation (ODE), such that:

$$\frac{dBA}{dt} = 1 + \alpha \cdot \left(\frac{1}{1 + e^{-r\,(t - t_{\text{start}})}} - \frac{2}{1 + e^{-r\,(t - t_{\text{end}})}} \right) + \varepsilon(t), \tag{2}$$

with initial condition $BA(t_0) = t_0$ (biological brain age equals chronological age). This ODE comprises two sigmoidal terms inducing nonlinear, age-dependent acceleration and deceleration of brain aging. Here, α controls the strength of brain aging, r is the transition rate, and ε introduces biological variability and measurement uncertainty [6]. To simulate AD-like aging trajectories, we introduce a gradual increase in brain age centred around t_{start} and then invoke dampening in later years, at t_{end} (where $t_{\text{start}} < t_{\text{end}}$), to model saturation of disease progression and to prevent the sampled brain ages from exceeding those the diffusion model generates. To produce AD-like trajectories where brain age surpasses chronological age (*e.g.*, by 5–10 years at age 80 [8,10]), we randomly sample parameters as $\alpha \sim \mathcal{N}(1.10, 0.05)$, $r \sim \mathcal{N}(0.25, 0.05)$, and $t_{\text{start}} \sim \mathcal{N}(55, 2.5)$. We also slightly alter the original healthy trajectories using this deviation model to simulate additional biological variability. Those parameters are sampled from $\alpha \sim \mathcal{N}(0.00, 0.10)$, $r \sim \mathcal{N}(0.25, 0.05)$, and $t_{\text{start}} \sim \mathcal{N}(55, 1.0)$, which are consistent with literature indicating negligible systematic deviation from chronological age [6]. Exemplary deviation functions are visualized in Fig. 2.

2.4 Experimental Setup

To evaluate the brain aging trajectories that we model, we build and classify INRs using *regular* temporal sampling (with a fixed number of scans and same

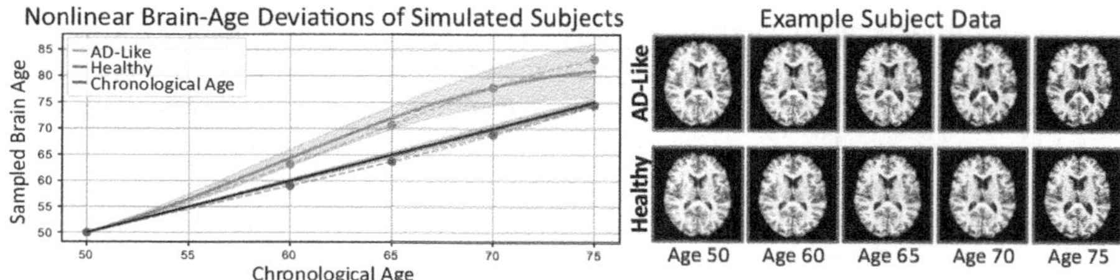

Fig. 2. Left: Exemplary simulated brain age deviation mappings (dashed lines) from our nonlinear ODE model. Solid lines denote the mean mapping and shaded regions indicate standard deviation across 450 simulations. Right: Visualization of sampled subject data as both healthy and AD-like trajectories.

acquisition times across all subjects) and *irregular* temporal sampling (with a different number of scans and acquisition times per subject). All experiments were conducted on a single Nvidia RTX4090 GPU with 24 GB. Code is available at https://github.com/AgamAulakh/INR-Trajectory-Classification.

Data: The diffusion model described in Sect. 2.3 is used to generate annual scans $\{I_0, \ldots, I_N\}$ of 450 simulated subjects at chronological ages $\{t_0, \ldots, t_N\}$, where $t_0 = 50$ and $t_N = 90$. This 3D MRI data is centre-cropped to dimensions of $147 \times 183 \times 169$ voxels with 1 mm isotropic voxel spacing. The 450 simulated subjects are divided into train and test sets following a 2:1 ratio, and for all subjects, the ages associated with each sampled scan are reassigned using the ODE model described in Sect. 2.3. Importantly, we include both trajectory pairs (healthy *and* AD-like) for subjects in the test set, but keep only one instance of a subject (healthy *or* AD-like) in the training data. Thus, although we do not train any classifier using paired trajectory data, the inclusion of both healthy and AD-like subject counterparts in the test set allows us to assess how different aging trajectories are interpreted as they act on the same brain morphology. With this simulated data, we consider a *regular* sampling scheme where, for all subjects, we select four scans at fixed chronological ages $\{50, 58, 67, 75\}$ for trajectory classification. We also consider a second, more realistic, *irregular* sampling scheme, where we randomly select three to five scans per subject within the range of 50–75 years.

INR Fitting: The INR architecture consists of eight residually connected layers of hidden size 512, with the spatial and temporal streams kept separate for the first five layers. The initialization INR θ^* is pretrained for 250 iterations, where we sample 0.1% of voxels from a batch size of three subjects per iteration. Then, for each subject, we finetune the initialization for 100 additional iterations to construct subject-specific INRs θ_m by sampling 1% of voxels. This process is followed for both sampling experiments independently. Initialization training takes less than one hour, and finetuning requires about 30 s per subject.

Classification: We train the combined encoder-classifier introduced in Sect. 2.2 for the regular and irregular sampling INRs independently. Processing the

INR parameters row-wise, the encoder sequentially expands the dimensionality through linear-batch norm-ReLU blocks of size 512, 1024, and 2048. The resulting latent representation after max pooling is then passed through two fully connected layers for classification. This model is trained for 100 epochs. To assess the advantages of our semi-disentangled INR architecture, we evaluate trajectory classification performance across single-stream (*e.g.*, $\mathbf{P}_{\text{space}}$) and multi-stream INR parameters (*e.g.*, $\mathbf{P}_{\text{time}} + \mathbf{P}_{\text{com}}$).

For initial baseline comparisons, we select a variant of the Simple Fully Convolutional Network (SFCN), which has been shown to achieve state-of-the-art performance for brain age prediction [18] and similarly strong results within the context of Alzheimer's disease classification [4]. We adapt the SFCN architecture used in [4] to process multiple scans simultaneously as multi-channel inputs. For the regular sampling experiment, the SFCN receives all four scans stacked in chronological order in the channel dimension. In the irregular sampling experiment, the SFCN receives the scans in chronological order with zero-padding applied if fewer than five scans are available.

3 Results and Discussion

We first assess the reconstructions produced by our semi-disentangled INR architecture when fit to *irregular* temporal data, and then present trajectory classification results across both temporal sampling experiments.

Quality of INR Reconstructions: Using our architecture (Sect. 2.4), we produce INRs with a maximum of 4.20 million parameters ($\mathbf{P}_{space} + \mathbf{P}_{time} + \mathbf{P}_{com}$), which results in data compression ranging from 18-30% for datasets with five to three scans each. Using subject-specific INRs trained with *irregular* temporal sampling, we reconstruct imaging data across all yearly timepoints included within our simulation. Table 1 summarizes the reconstruction data quality across timepoints used for INR finetuning (training data) and those not included (interpolated/extrapolated timepoints). We observe lower quality for reconstructions generated through extrapolations, which may indicate that the INRs default to producing the average trajectory learned during initialization for timepoints where prior or subsequent subject data is not available. Reconstructions from INRs of two subjects are shown in Fig. 3, demonstrating that the INRs can

Table 1. Quality of reconstructed training data, interpolations, and extrapolations from INRs in the *irregular* sampling experiment. We report Mean Square Error (MSE), Peak Signal-to-Noise Ratio (PSNR), and Structural Similarity Index Measure (SSIM).

Reconstruction	Healthy Subjects			AD-Like Subjects		
	MSE	PSNR	SSIM	MSE	PSNR	SSIM
Training Data	3.95×10^{-3}	30.2	0.953	3.91×10^{-3}	30.2	0.952
Interpolation	5.00×10^{-3}	29.3	0.943	6.63×10^{-3}	28.3	0.930
Extrapolation	10.6×10^{-3}	26.1	0.898	12.1×10^{-3}	25.6	0.887

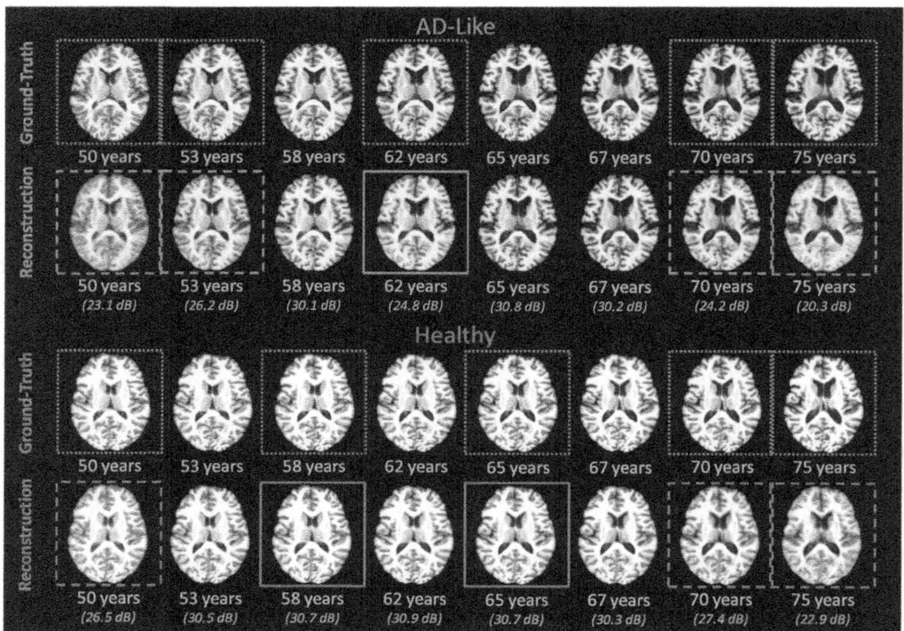

Fig. 3. Examples of reconstructed scans for simulated subjects with AD-like (orange) and healthy (blue) brain aging trajectories. For both AD-like and healthy trajectories, the top row shows the ground-truth scans and the bottom row shows their corresponding reconstructions. Grey dotted outlines indicate data not used for training. Solid outlines (orange/blue) indicate interpolated reconstructions and dashed outlines (orange/blue) indicate extrapolated reconstructions. (Color figure online)

maintain subject-specific morphology and produce a stable trajectory across the temporal range without closely overfitting to (*i.e.*, only producing viable reconstructions of) timepoints used for finetuning.

Brain Aging Trajectory Classification: Table 2 shows classification results across *regular* and *irregular* sampling experiments. We achieve consistently strong classification performance using $\mathbf{P}_{\text{time}}$ and $\mathbf{P}_{\text{time}} + \mathbf{P}_{\text{com}}$ INR parameters. In particular, these two parameter sets outperform the SFCN baseline in the irregular sampling scheme, which highlights the advantages of encoding longitudinal data as continuous representations. Furthermore, we observe chance-level classification accuracies using $\mathbf{P}_{\text{space}}$ across both sampling schemes, and show that excluding these parameters leads to notable improvements (*e.g.*, 77% when using $\mathbf{P}_{\text{space}} + \mathbf{P}_{\text{time}} + \mathbf{P}_{\text{com}}$ *vs.* 81% with $\mathbf{P}_{\text{time}}$). This highlights the benefits of our semi-disentangled INR architecture, as we can selectively exclude parameters which may introduce spurious correlations that do not generalize beyond training data and, as a result, construct parameter-efficient representations. While we observe the highest accuracies using $\mathbf{P}_{\text{time}}$, different parameter combinations may be useful for other scenarios where, for example, one is interested in studying brain dynamics with acute and spatially-localized temporal changes.

Table 2. Classification results across single- and multi-stream INR parameters with SFCN benchmark. Number of parameters used for classification is reported in the order of millions (M). $\mathbf{P}_{\text{space}}$, $\mathbf{P}_{\text{time}}$, and $\mathbf{P}_{\text{com}}$ are abbreviated as $\mathbf{P}_{\text{s}}$, $\mathbf{P}_{\text{t}}$, and $\mathbf{P}_{\text{c}}$, respectively.

Model	Regular Sampling		Irregular Sampling	
	Input Size (↓)	Accuracy (↑)	Input Size (↓)	Accuracy (↑)
SFCN	18.1M	**97.7**	22.7M	73.7
INR ($\mathbf{P}_s$)	2.10M	50.0	2.10M	49.8
INR ($\mathbf{P}_t$)	**1.05M**	95.6	**1.05M**	**81.3**
INR ($\mathbf{P}_s + \mathbf{P}_c$)	3.15M	91.7	3.15M	74.6
INR ($\mathbf{P}_t + \mathbf{P}_c$)	2.10M	96.3	2.10M	78.3
INR ($\mathbf{P}_s + \mathbf{P}_t + \mathbf{P}_c$)	4.20M	95.3	4.20M	77.3

To keep comparisons consistent, we do not provide the INR encoder and the SFCN with temporal information associated with the imaging data. However, it is implicitly encoded in the INR parameters during finetuning. This may be one reason why the SFCN performs poorly with irregular sampling, and, therefore, our use of this model may not provide the most competitive benchmark in this initial evaluation. Despite this, our INR-based results are highly promising and motivate immediate future work to use real data with more advanced baselines.

4 Conclusion

This paper presents a novel method for modelling and, for the first time, classifying brain aging trajectories using spatiotemporal INRs. We introduce a new INR architecture capable of partially disentangling spatiotemporal parameters and highlight its advantages in isolating informative features for the classification of healthy and AD-like brain aging trajectories. Using a newly designed trajectory simulation framework, we demonstrate that our proposed INR architecture can represent high-resolution 3D longitudinal MRI data with both regular and irregular temporal sampling, capture structural brain aging dynamics, and be directly applied in a downstream classification task. A notable strength of our approach is the flexibility it offers, both in adapting to data with varying sampling patterns and in its capacity to operate directly on neuroimaging data to disentangle spatial and temporal characteristics. Extending the analysis provided in this paper, future work is required to test this approach on real data.

Acknowledgments. This work was supported by the Department of Pediatrics and the Azrieli Accelerator at the University of Calgary, the Hotchkiss Brain Institute, the Alberta Children's Hospital Foundation, and the Natural Sciences and Engineering Research Council of Canada (RGPIN-03532-2023).

Disclosure of Interests. The authors have no competing interests to declare.

References

1. Bauer, M., Dupont, E., Brock, A., Rosenbaum, D., Schwarz, J.R., Kim, H.: Spatial functa: scaling functa to imagenet classification and generation (2023)
2. Cui, R., Liu, M., Initiative, A.D.N., et al.: RNN-based longitudinal analysis for diagnosis of Alzheimer's disease. Comput. Med. Imaging Graph. **73**, 1–10 (2019)
3. Dupont, E., Kim, H., Eslami, S., Rezende, D., Rosenbaum, D.: From data to Functa: your data point is a function and you can treat it like one. arXiv preprint arXiv:2201.12204 (2022)
4. Early, S.E., Wilms, M., Forkert, N.D.: A comparison of biomarker modalities for predicting disease progression in dementia patients. In: Medical Imaging 2025: Imaging Informatics, vol. 13411, pp. 27–33. SPIE (2025)
5. Essakine, A., et al.: Where do we stand with implicit neural representations? a technical and performance survey. arXiv preprint arXiv:2411.03688 (2024)
6. Franke, K., Gaser, C.: Ten years of brainage as a neuroimaging biomarker of brain aging: what insights have we gained? Front. Neurol. **10**, 789 (2019)
7. Friedrich, P., Bieder, F., Cattin, P.C.: Medfuncta: modality-agnostic representations based on efficient neural fields. arXiv preprint arXiv:2502.14401 (2025)
8. Gaser, C., Franke, K., Klöppel, S., Koutsouleris, N., Sauer, H.: Brainage in mild cognitive impaired patients: predicting the conversion to Alzheimer's disease. PLoS ONE **8**(6), e67346 (2013)
9. Gielisse, A., van Gemert, J.: End-to-end implicit neural representations for classification. In: Proceedings of the Computer Vision and Pattern Recognition Conference, pp. 18728–18737 (2025)
10. Habes, M., et al.: The brain chart of aging: machine-learning analytics reveals links between brain aging, white matter disease, amyloid burden, and cognition in the istaging consortium of 10,216 harmonized mr scans. Alzheimer's Dementia **17**(1), 89–102 (2021)
11. Hecht-Nielsen, R.: On the algebraic structure of feedforward network weight spaces. In: Advanced Neural Computers, pp. 129–135. Elsevier (1990)
12. Luigi, L.D., Cardace, A., Spezialetti, R., Ramirez, P.Z., Salti, S., di Stefano, L.: Deep learning on implicit neural representations of shapes. In: The Eleventh International Conference on Learning Representations (2023)
13. Mofrad, S.A., Lundervold, A.J., Vik, A., Lundervold, A.S.: Cognitive and MRI trajectories for prediction of Alzheimer's disease. Sci. Rep. **11**(1), 2122 (2021)
14. Mueller, S.G., et al.: The Alzheimer's disease neuroimaging initiative. Neuroimaging Clin. **15**(4), 869–877 (2005)
15. Ouyang, J., et al.: Self-supervised longitudinal neighbourhood embedding. In: de Bruijne, M., et al. (eds.) MICCAI 2021. LNCS, vol. 12902, pp. 80–89. Springer, Cham (2021). https://doi.org/10.1007/978-3-030-87196-3_8
16. Ouyang, J., Zhao, Q., Adeli, E., Zaharchuk, G., Pohl, K.M.: Self-supervised learning of neighborhood embedding for longitudinal MRI. Med. Image Anal. **82**, 102571 (2022)
17. Ouyang, J., et al.: Longitudinal pooling & consistency regularization to model disease progression from MRIs. IEEE J. Biomed. Health Inform. **25**(6), 2082–2092 (2020)

18. Peng, H., Gong, W., Beckmann, C.F., Vedaldi, A., Smith, S.M.: Accurate brain age prediction with lightweight deep neural networks. Med. Image Anal. **68**, 101871 (2021)
19. Poulakis, K., et al.: Multi-cohort and longitudinal Bayesian clustering study of stage and subtype in Alzheimer's disease. Nat. Commun. **13**(1), 4566 (2022)
20. Rahaman, N., et al.: On the spectral bias of neural networks. In: International Conference on Machine Learning, pp. 5301–5310. PMLR (2019)
21. Saragadam, V., LeJeune, D., Tan, J., Balakrishnan, G., Veeraraghavan, A., Baraniuk, R.G.: Wire: wavelet implicit neural representations. In: Proceedings of the IEEE/CVF Conference on Computer Vision and Pattern Recognition, pp. 18507–18516 (2023)
22. Shuaibu, A.L., Gibb, K.A., Wijeratne, P.A., Simpson, I.J.: Capturing longitudinal changes in brain morphology using temporally parameterized neural displacement fields. arXiv preprint arXiv:2504.09514 (2025)
23. Ulrich, C., Zimmerer, D., Stiefelhagen, R.: Back to the future: challenges of sparse and irregular medical image time series. In: Schroder, A., et al. (eds.) MICCAI 2024 . LNCS, vol. 15401, p. 15. Springer, Cham (2025). https://doi.org/10.1007/978-3-031-84525-3_2
24. Wilms, M., et al.: A lightweight 3d conditional diffusion model for self-explainable brain age prediction in adults and children. In: Bathula, D.R., et al. (eds.) MLCN 2024. LNCA, vol. 15266, pp. 57–67. Springer, Cham (2025). https://doi.org/10.1007/978-3-031-78761-4_6
25. Wolterink, J.M., Zwienenberg, J.C., Brune, C.: Implicit neural representations for deformable image registration. In: International Conference on Medical Imaging with Deep Learning, pp. 1349–1359. PMLR (2022)

Improving Virtual Contrast Enhancement Using Longitudinal Data

Pierre Fayolle[1,2,5](✉), Alexandre Bône[1], Noëlie Debs[1], Philippe Robert[1], Pascal Bourdon[4,5], Rémy Guillevin[2,3,5], and David Helbert[4,5]

[1] Guerbet Research, Villepinte, France
pierre.fayolle@univ-poitiers.fr
[2] LMA, Université de Poitiers, Poitiers, France
[3] CHU de Poitiers, Poitiers, France
[4] XLIM, Université de Poitiers, Poitiers, France
[5] I3M, Common Laboratory CNRS-Siemens, Poitiers, France

Abstract. Gadolinium-based contrast agents (GBCAs) are widely used in magnetic resonance imaging (MRI) to enhance lesion detection and characterisation, particularly in the field of neuro-oncology. Nevertheless, concerns regarding gadolinium retention and accumulation in brain and body tissues, most notably for diseases that require close monitoring and frequent GBCA injection, have led to the need for strategies to reduce dosage.

In this study, a deep learning framework is proposed for the virtual contrast enhancement of full-dose post-contrast T1-weighted MRI images from corresponding low-dose acquisitions. The contribution of the presented model is its utilisation of longitudinal information, which is achieved by incorporating a prior full-dose MRI examination from the same patient. A comparative evaluation against a non-longitudinal single session model demonstrated that the longitudinal approach significantly improves image quality across multiple reconstruction metrics. Furthermore, experiments with varying simulated contrast doses confirmed the robustness of the proposed method. These results emphasize the potential of integrating prior imaging history into deep learning-based virtual contrast enhancement pipelines to reduce GBCA usage without compromising diagnostic utility, thus paving the way for safer, more sustainable longitudinal monitoring in clinical MRI practice.

Keywords: Magnetic Resonance Imaging (MRI) · Gadolinium-Based Contrast Agents (GBCA) · Contrast Dose Reduction · Longitudinal Imaging · Deep Learning

1 Introduction

Magnetic Resonance Imaging (MRI) is an essential diagnostic tool in clinical practice, offering non-invasive, high-resolution visualization of anatomical and

B. Hou and T. S. Mathai (Eds.): LMID 2025, LNCS 16184, pp. 46–55, 2026.
https://doi.org/10.1007/978-3-032-16128-4_5

functional structures. Contrast-enhanced MRI, particularly T1-weighted imaging following the administration of gadolinium-based contrast agents (GBCAs), has been shown to significantly improve lesion detection and characterisation, especially in the fields of neuroimaging and oncology [18]. However, concerns have been raised regarding the long-term safety of GBCAs, including gadolinium retention and accumulation in brain and body tissues, even in patients with normal renal function [4,8]. These issues are of particular significance for patients afflicted with brain tumours or other chronic conditions, who require regular follow-up examinations involving contrast-enhanced MRI.

In order to address the aforementioned risks, a critical research objective has emerged: the reduction of the dose of contrast agent without compromising image quality. However, lower doses typically result in diminished signal enhancement and impaired diagnostic performance [3]. Recently, deep learning–based approaches have emerged as powerful tools for virtually enhancing medical images from low-dose acquisitions or simulations [1,3,5,10,11,17]. These techniques have been demonstrated to be capable of effectively learning spatial and contrast relationships in paired data, with the potential to significantly reduce contrast agent usage. Although synthesized contrast-enhanced images appear realistic, reader studies have revealed critical limitations, including missed subtle lesions, hallucinated findings [1,5,10], and excessive smoothing [1], which prevent their use in clinical routine.

A fundamental innovation of our approach is the utilisation of not only the current low-dose MRI, but also the patient's previous MRI exam acquired at full contrast dose. The model is trained on paired MRI volumes composed of pre- and post-contrast images from two sequential MRI sessions: one with a standard 100% dose (Session 1) and another with an ajustable synthetic low-dose (Session 2). The longitudinal design of the model might facilitate the acquisition of subject-specific contrast enhancement patterns and anatomical knowledge, thereby enhancing its capacity to synthesize high-quality full-dose contrast images. The objective of this study is to evaluate whether incorporating prior clinical imaging data through a deep learning–based virtual contrast enhancement approach leads to improved image fidelity compared to conventional non-longitudinal methods, thereby supporting the feasibility of reducing GBCA usage.

2 Materials and Methods

2.1 Longitudinal Virtual Contrast Enhancement Workflow

Figure 1 shows an overview of the proposed longitudinal virtual contrast enhancement method for standard-dose post-contrast T1-weighted (T1-SD) MRI images. This method uses prior session data and low-dose acquisitions from a follow-up session. The diagram is divided into two temporal phases: the previous session (ses-01) and the follow-up session (ses-02). In the previous session, a pre-contrast T1-weighted image (T1-PC) is acquired prior to the administration of

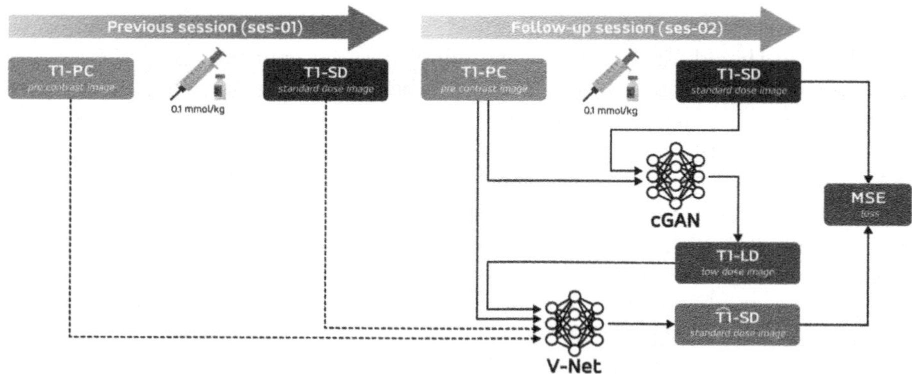

Fig. 1. Workflow of the Proposed Longitudinal Virtual Contrast Enhancement Model for T1-Weighted Images. Dotted arrow shows the main contribution of the proposed work. cGAN: conditional Generative Adversarial Networks, MSE: Mean Square Error.

0.1 mmol/kg gadolinium contrast agent, followed by the acquisition of T1-SD. These images thus serve as high-quality historical references. In the follow-up session (ses-02), a new T1-PC image is acquired and a standard dose is administered in a manner similar to that of ses-01. However, in the proposed framework, a simulation of a low-dose equivalent (T1-LD) from the follow-up T1-PC and T1-SD is conducted using a conditional generative adversarial network (cGAN). This synthetic low-dose images are generated by simulating contrast levels corresponding to 10%, 15%, 20%, 25% and 33% of the standard gadolinium dose. The T1-PC and T1-LD from ses-02, along with the T1-PC and T1-SD from ses-01, are then subsequently concatenated and provided as multi-channel inputs to a 3D V-Net architecture. This network is trained to generate an image that approximates the full-dose post-contrast image $\widehat{\text{T1}}$-SD for the follow-up session. The predicted image is then compared to the actual follow-up T1-SD using a mean squared error (MSE) loss, enforcing voxel-wise intensity fidelity.

2.2 Public Dataset

This study uses images from the ACRIN-DSC-MR-Brain collection [9], a publicly available dataset hosted on The Cancer Imaging Archive (TCIA) at the following link: https://www.cancerimagingarchive.net/collection/acrin-dsc-mr-brain/. This longitudinal dataset includes 123 patients diagnosed with recurrent glioblastoma undergoing anti-angiogenic therapy. For this work, only the axial pre- and post-contrast T1-weighted images from the two most recent imaging sessions available for each subject were used. A total of 26 patients were excluded from the dataset due to the absence of suitable pre- and post-contrast T1-weighted images from two separate sessions. Of the remaining 97 patients left, the average time between two sessions is 57.21 ± 31.72 days, ranging from a minimum of 16 days to a maximum of 312 days. The dataset was split into 75 subjects for training, 5 for validation, and 17 for testing.

2.3 Data Preprocessing

The two sessions for each subject were processed sequentially. For both T1-PC and T1-SD images, skull stripping was performed using HD-BET [7]. Prior to that, all volumes were cropped to the brain region and resampled to isotropic $1\,\text{mm}^3$ resolution. The T1-SD image from the ses-01 was elastically registered to the corresponding T1-PC image using SimpleITK extension SimpleElastix [12], and both were min-max normalized jointly to preserve relative intensity scales. For the ses-02, T1-PC and T1-SD images were first aligned to the T1-PC from the ses-01 using the same registration method to ensure inter-session correspondence. The same crop box from the ses-01 was applied to the ses-02 volumes for consistent spatial coverage. As the ACRIN dataset does not provide T1-LD, a synthetic contrast reduction approach was used, as proposed in [16], to simulate adjustable dose T1ce images from the ses-02. Subsequently, all images from the ses-02 (T1-PC, T1-SD, and T1-LD) were jointly normalized using the identical strategy employed in the preceding session. Following these preprocessing steps, the four resulting volumes, T1-PC and T1-SD from the ses-01, with T1-PC and T1-LD from the ses-02, were stacked along the channel dimension to form an input of size $160 \times 192 \times 160 \times 4$.

To improve the model generalization and to keep the input dimensions consistent, a complete data augmentation pipeline was used during training. Spatial variability was introduced through random 3D flips along each anatomical axis (applied with a probability of 0.5 per axis) and random affine transformations, including small rotations (up to 0.05 rad), translations (up to 5 voxels), and uniform scaling within a $\pm 10\%$ range. In addition to spatial augmentations, intensity-based transformations were applied to increase robustness to signal variations. These included the addition of random Gaussian noise (standard deviation of 0.01, applied with 30% probability) and random intensity shifts (offset of 0.1, applied with 50% probability).

2.4 Training Strategy

We trained a 3D V-Net model proposed in [2] and originally adapted from [13] on whole-brain volumes. The network was trained for 500 epochs with a batch size of 1 and an initial learning rate of 10^{-4}. Optimization was performed using the Adam optimizer. To adapt the learning rate during training, we employed a scheduler that reduced the learning rate by a factor of 0.5 when the loss stabilized for 10 consecutive epochs. This dynamic adjustment encouraged more stable convergence throughout training.

2.5 Model Evaluation

To assess the effectiveness of the proposed longitudinal model, its performance was compared to that of a single session model that was trained only on data

Table 1. Quantitative Results Between Low-Dose T1 (T1-LD) Images, Single Session and Longitudinal Models Across Reconstruction Metrics Compared with Full-Dose T1 (ses-02 T1-SD) Ground-Truth.

Model	MSE ($\times 10^{-2}$) (↓)	PSNR (dB) (↑)	SSIM (↑)
T1-LD	0.2211 ± 0.2292	28.2346 ± 3.7540	0.9315 ± 0.0223
Single Session	0.1564 ± 0.1190	29.1920 ± 3.4279	0.9317 ± 0.0166
Longitudinal	**0.1160** ± 0.1555	**32.6337** ± 5.9517	**0.9726** ± 0.0177
	$p = 0.0569$*	$p = 0.0083$*	$p < 0.0001$*

*The reported p-values indicate the statistical significance of improvements in the longitudinal model compared to the single session baseline.

from the ses-02, called the single session model. This approach aligns with the strategies commonly adopted in recent studies aimed at reducing contrast doses [3,6,14,15,17]. Importantly, both the models were trained using the same architecture, loss functions, and hyperparameters, ensuring a fair comparison. For that purpose, three different reconstruction metrics were used. The Mean Square Error (MSE, $\times 10^{-2}$), the Peak Signal-to-Noise Ratio (PSNR, dB) and the SSIM (Structural SIMilarity). Appropriate paired tests were conducted for each metric to assess the statistical significance of the difference between the two models. This comparative evaluation provides a clearer insight into the impact of longitudinal information on the virtual contrast enhancement quality. Each training was performed on a NVIDIA Tesla T4 with 16Gb of RAM, taking approximatively 23 h for the single session model and 25 h for the longitudinal model. Code will be shared upon acceptance of our work.

3 Results

3.1 Comparative Quantitative Analysis

Table 1 shows the quantitative results of the single session model compared to the longitudinal model at 25% dose. For the single session model, the MSE was 0.1564 ± 0.1190, PSNR was 29.1920 ± 3.4279, and the SSIM was 0.9317 ± 0.0166. In contrast, the longitudinal model yielded improved performance with an MSE of 0.1160 ± 0.1555, PSNR of 32.6337 ± 5.9517, and SSIM of 0.9726 ± 0.0177.

For MSE, a Shapiro-Wilk test indicated non-normality ($p = 0.0018$), and the Wilcoxon signed-rank test was therefore used, revealing a trend toward significance ($p = 0.0569$). For PSNR and SSIM, normality was confirmed ($p = 0.9645$ and $p = 0.3069$, respectively), and paired t-tests were applied. The PSNR comparison yielded a statistically significant improvement ($t = -3.01$, $p = 0.0083$), and SSIM showed a highly significant difference ($t = -11.73$, $p = 2.85 \times 10^{-9}$). These results support the conclusion that incorporating longitudinal information substantially improves image quality across multiple quantitative metrics.

To visually illustrate these differences, Fig. 2 presents boxplots comparing the two models across the three metrics. The distributions clearly highlight the

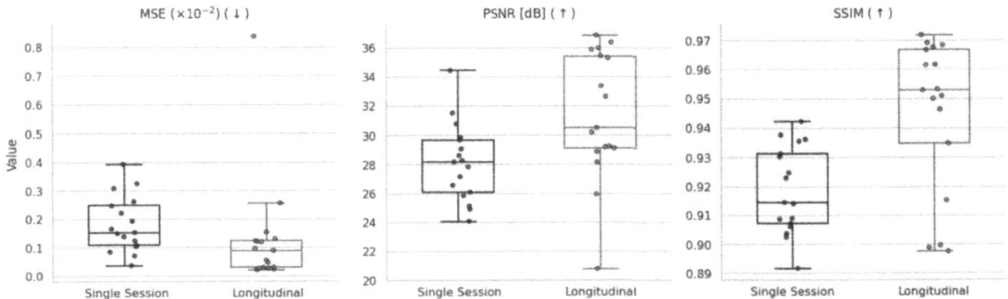

Fig. 2. Boxplot Analysis of Metrics Between the Single Session and Longitudinal Models. MSE: Mean Square Error, PSNR: Peak Signal-to-Noise Ratio, SSIM: Structural Similarity Index.

reduced variance and overall improvement in performance when longitudinal information is integrated.

3.2 Qualitative Comparison

As illustrated in Fig. 3, representative lesion-centered slices from two distinct test subjects are presented for a 25% synthetized dose, thereby enabling a qualitative comparison between the single session and the proposed longitudinal models. In all cases, both methods were found to be equally successful in virtually enhancing contrast in anatomically coherent images, with consistent lesion localisation and structural delineation. It is noteworthy that visual inspection reveals only subtle differences between the two approaches, particularly in regions of lesion enhancement and surrounding tissue contrast. These observations are consistent with the robust performance of both models and corroborate the quantitative findings, wherein the longitudinal model demonstrates statistically significant improvements despite minimal perceptible differences in visual appearance.

3.3 Dose Variation Evaluation

Figure 4 shows the virtual contrast enhancement performance for both the single session and longitudinal approaches at different dose levels (10%, 15%, 20%, 25% and 33%). Interestingly, at a dose of 15%, both models converge in three different reconstruction metrics. For other dose levels, however, the longitudinal model consistently outperforms the single session model in all metrics. The regression slopes are not statistically significant ($p > 0.05$), indicating stable performance across doses.

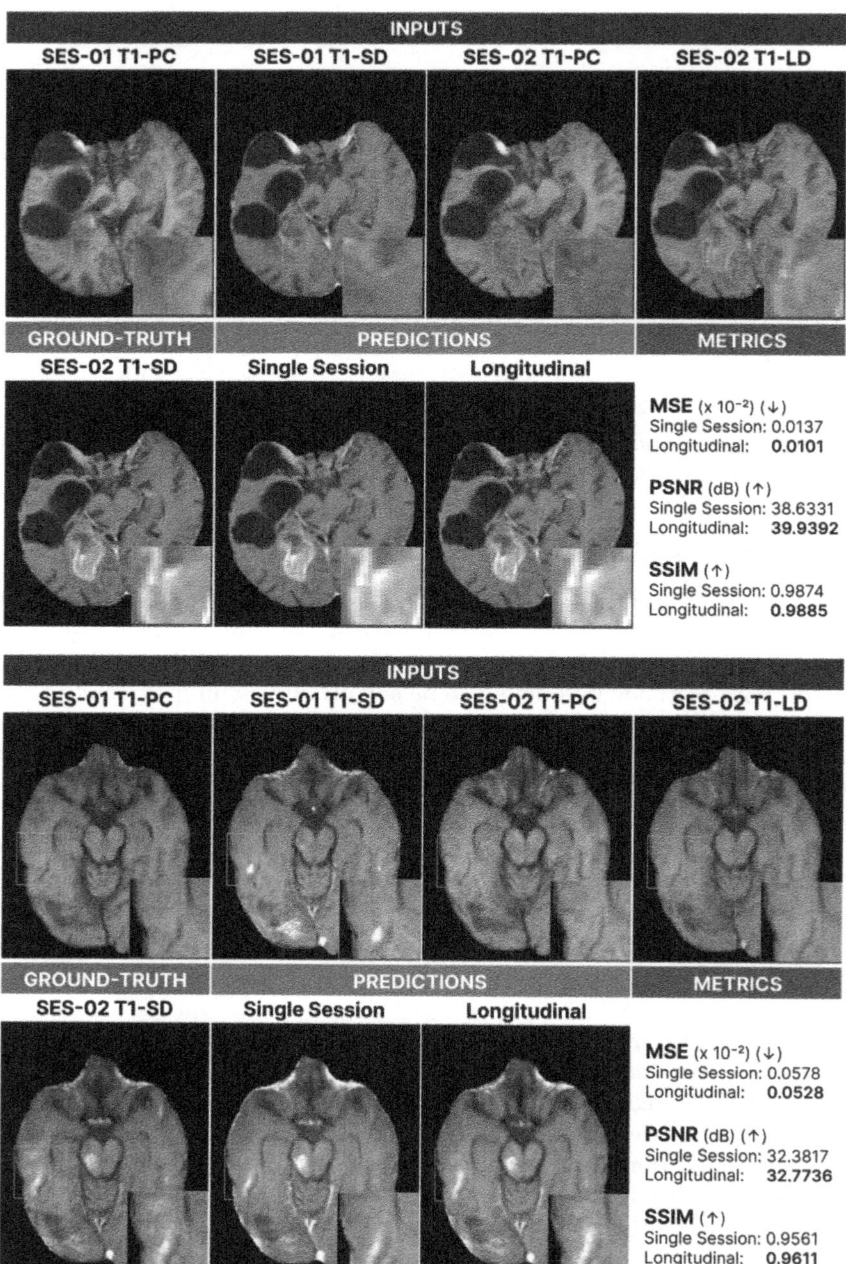

Fig. 3. Comparison Between Single Session and Longitudinal Data-Driven Models. PC: Pre-contrast, SD: Standard-dose, LD: Low-dose (25% synthetized dose).

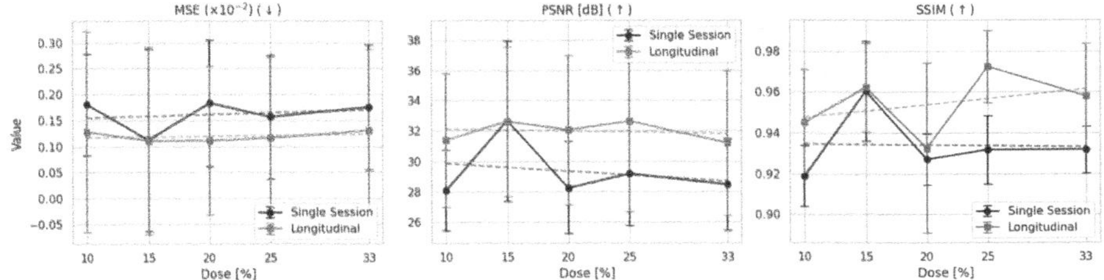

Fig. 4. Effect of Simulated Dose Levels on the Virtual Contrast Enhancement Metrics. Dotted lines represent linear regression curves fitted across dose levels for each model. Error bars denote standard deviation.

4 Discussions

In this study, a longitudinal data-driven model was proposed with the objective of improving the performance of virtual contrast enhancement MRI models for patients undergoing multiple follow-up examinations. The performance of the presented model was quantitatively compared to that of a single session model that had been trained exclusively on second-session images. The longitudinal model demonstrated significant improvements in image fidelity and structural similarity across a range of reconstruction metrics. The findings emphasise the efficacy of incorporating longitudinal data for virtual contrast enhancement tasks and the potential for reducing contrast dose.

In this work, only T1-weighted sequences were utilised to train the model. Although prior non-longitudinal studies have investigated the incorporation of supplementary MRI modalities, including T2-FLAIR and ADC [1,2], these studies did not demonstrate a significant improvement in performance with the incorporation of these sequences [2]. Nevertheless, the integration and evaluation of such additional modalities in the context of longitudinal analyses may warrant further investigation.

As the public dataset did not include real low-dose images, synthetic low-dose inputs were generated based on the method described in [16]. The utilization of synthetic data offers a certain degree of flexibility, allowing for controlled variation of dose levels to simulate a wide range of conditions and to augment the dataset. This approach would not be feasible within the constraints of clinical protocols in medical research. However, it should be noted that these synthesized images do not fully replicate the complexity and variability of actual low-dose acquisitions, and several factors could limit the reliability of the results obtained using them. In order to ensure clinical relevance and robustness, further validation should be performed with real-dose images, particularly in the context of clinical deployment.

While the proposed model demonstrates promising robustness across a wide range of time intervals (16–312 days), significant anatomical and perfusion changes, particularly between pre- and postoperative scans, can restrict the effectiveness of longitudinal data. Such variability may affect the model's ability to leverage prior data effectively. Although the framework was designed to handle moderate inter-session changes, pre- and post-operative paired data were not experimented with in this study. Its performance under such conditions remains to be investigated.

It is important to note that the test set utilised in this study comprises only 17 patients, all obtained from a single publicly available dataset. Despite its limited size, it includes patients with treated lesions that demonstrate a range of longitudinal responses, such as growth, shrinkage, and stability. This heterogeneity introduces valuable clinical variability, thereby enabling the model to be evaluated in a more realistic setting. Nevertheless, the restricted number and source of test cases may limit the generalizability of the findings. It is therefore recommended that future research include larger, multi-institutional cohorts with the aim of validating the proposed approach further across a range of lesion types and treatment responses.

5 Conclusion

The present study proposes a proof-of-concept longitudinal deep learning approach for virtually enhancing contrast of low-dose T1-weighted MRI acquisitions to generate full-dose post-contrast images. This approach utilises prior high-dose scans to enhance image fidelity and has been demonstrated to significantly improve performance in comparison to a conventional non-longitudinal single session model across a range of quantitative reconstruction metrics. Furthermore, the approach demonstrates robustness across varying synthetised contrast levels, thereby confirming its adaptability to different clinical scenarios. These findings emphasize the potential of incorporating longitudinal information into contrast dose reduction strategies and demonstrate the feasibility of safer, lower-dose MRI protocols without compromising diagnostic quality. Clinically speaking, patients with chronic diseases who require regular and frequent gadolinium injections represent a key population for whom implementing protocols with alternating full and low-dose acquisitions may offer a favorable risk-benefit profile.

Acknowledgments. This work was granted access to the HPC resources of IDRIS under the allocation 2024-A0150314655 made by GENCI. This work was supported by ANRT (CIFRE 2023/1206).

Disclosure of Interests. P.F., A.B., N.D., P.R. are employees of Guerbet.

References

1. Ammari, S., et al.: Can deep learning replace gadolinium in neuro-oncology?: a reader study. Invest. Radiol. **57**(2), 99–107 (2022)
2. Bône, A., et al.: Contrast-enhanced brain MRI synthesis with deep learning: key input modalities and asymptotic performance. In: 2021 IEEE 18th International Symposium on Biomedical Imaging (ISBI), pp. 1159–1163. IEEE (2021)

3. Gong, E., Pauly, J.M., Wintermark, M., Zaharchuk, G.: Deep learning enables reduced gadolinium dose for contrast-enhanced brain MRI. J. Magn. Reson. Imaging **48**(2), 330–340 (2018)
4. Gulani, V., Calamante, F., Shellock, F.G., Kanal, E., Reeder, S.B.: Gadolinium deposition in the brain: summary of evidence and recommendations. Lancet Neurol. **16**(7), 564–570 (2017)
5. Haase, R., et al.: Reduction of gadolinium-based contrast agents in MRI using convolutional neural networks and different input protocols: limited interchangeability of synthesized sequences with original full-dose images despite excellent quantitative performance. Invest. Radiol. **58**(6), 420–430 (2023)
6. Haase, R., et al.: Artificial contrast: deep learning for reducing gadolinium-based contrast agents in neuroradiology. Invest. Radiol. **58**(8), 539–547 (2023)
7. Isensee, F., et al.: Automated brain extraction of multisequence MRI using artificial neural networks. Hum. Brain Mapp. **40**(17), 4952–4964 (2019)
8. Kanda, T., Ishii, K., Kawaguchi, H., Kitajima, K., Takenaka, D.: High signal intensity in the dentate nucleus and globus pallidus on unenhanced T1-weighted MR images: relationship with increasing cumulative dose of a gadolinium-based contrast material. Radiology **270**(3), 834–841 (2014)
9. Kinahan, P., Muzi, M., Bialecki, B., Herman, B., Coombs, L.: Data from ACRIN-DSC-MR-BRAIN. Cancer Imaging Arch. (2019)
10. Luo, H., et al.: Deep learning-based methods may minimize GBCA dosage in brain MRI. Eur. Radiol. **31**, 6419–6428 (2021)
11. Mallio, C.A., et al.: Artificial intelligence to reduce or eliminate the need for gadolinium-based contrast agents in brain and cardiac MRI: a literature review. Invest. Radiol. **58**(10), 746–753 (2023)
12. Marstal, K., Berendsen, F., Staring, M., Klein, S.: SimpleElastix: a user-friendly, multi-lingual library for medical image registration. In: Proceedings of the IEEE Conference on Computer Vision and Pattern Recognition Workshops, pp. 134–142 (2016)
13. Milletari, F., Navab, N., Ahmadi, S.A.: V-net: fully convolutional neural networks for volumetric medical image segmentation. In: 2016 Fourth International Conference on 3D Vision (3DV), pp. 565–571. IEEE (2016)
14. Mingo, A.F., et al.: Amplifying the effects of contrast agents on magnetic resonance images using a deep learning method trained on synthetic data. Invest. Radiol. **58**(12), 853–864 (2023)
15. Pasumarthi, S., Tamir, J.I., Christensen, S., Zaharchuk, G., Zhang, T., Gong, E.: A generic deep learning model for reduced gadolinium dose in contrast-enhanced brain MRI. Magn. Reson. Med. **86**(3), 1687–1700 (2021)
16. Pinetz, T., Kobler, E., Haase, R., Deike-Hofmann, K., Radbruch, A., Effland, A.: Faithful synthesis of low-dose contrast-enhanced brain MRI scans using noise-preserving conditional GANs. In: Greenspan, H., et al. (eds.) MICCAI 2023. LNCS, vol. 14221, pp. 607–617. Springer, Cham (2023). https://doi.org/10.1007/978-3-031-43895-0_57
17. Pinetz, T., et al.: Gadolinium dose reduction for brain MRI using conditional deep learning. arXiv preprint arXiv:2403.03539 (2024)
18. Runge, V.M.: Safety of approved MR contrast media for intravenous injection. J. Magn. Reson. Imaging: Off. J. Int. Soc. Magn. Reson. Med. **12**(2), 205–213 (2000)

Spatio-Temporal Conditional Diffusion Models for Forecasting Future Multiple Sclerosis Lesion Masks Conditioned on Treatments

Gian Mario Favero[1,2(✉)], Ge Ya Luo[2], Nima Fathi[1,2], Justin Szeto[1,2], Douglas L. Arnold[1], Brennan Nichyporuk[1,2], Chris Pal[2], and Tal Arbel[1,2]

[1] McGill University, Montreal, Canada
[2] Mila – Quebec AI Institute, Montreal, Canada
gian.favero@mail.mcgill.ca

Abstract. Image-based personalized medicine has the potential to transform healthcare, particularly for diseases that exhibit heterogeneous progression such as Multiple Sclerosis (MS). In this work, we introduce the first treatment-aware spatio-temporal diffusion model that is able to generate *future* masks demonstrating lesion evolution in MS. Our voxel-space approach incorporates multi-modal patient data, including MRI and treatment information, to forecast new and enlarging T2 (NET2) lesion masks at a future time point. Extensive experiments on a multi-centre dataset of 2131 patient 3D MRIs from randomized clinical trials for relapsing-remitting MS demonstrate that our generative model is able to accurately predict NET2 lesion masks for patients across six different treatments. Moreover, we demonstrate our model has the potential for real-world clinical applications through downstream tasks such as future lesion count and location estimation, binary lesion activity classification, and generating counterfactual future NET2 masks for several treatments with different efficacies. This work highlights the potential of causal, image-based generative models as powerful tools for advancing data-driven prognostics in MS.

Keywords: Diffusion · Multiple Sclerosis · Spatio-Temporal

1 Introduction

Deep learning models for image-based personalized medicine that predict future individual patient outcomes enable early and more informed treatment interventions before irreversible disease accrual occurs. This potential is particularly important for diseases such as Multiple Sclerosis (MS) and certain cancers, which are characterized by heterogeneous evolutions and variable treatment effects. Recent work has shown success in leveraging baseline MRI to forecast patient outcomes in MS [4,5]. However, beyond predicting numerical outcomes, a crucial question remains: What if we could also predict personalized future images

B. Hou and T. S. Mathai (Eds.): LMID 2025, LNCS 16184, pp. 56–67, 2026.
https://doi.org/10.1007/978-3-032-16128-4_6

depicting the appearance of future pathological structure changes (e.g., new and enlarging lesions or tumor metastases) on a patient's brain MRI? Such visual forecasts would not only increase the trustworthiness of predictions through enhanced explainability, but would also provide deeper insights into the effects of various treatments at the individual level, thereby improving clinical decision support.

In the context of MS, lesions, visible as hyperintense regions on T2-weighted MRI, serve as key markers for tracking disease activity. New and enlarging T2 (NET2) lesions are especially important for assessing disease progression and treatment efficacy in relapsing-remitting MS (RRMS) [3,22]. However, due to their extremely small size (typically less than 0.1% of brain tissue) and irregular distribution with lesions predominantly forming in regions such as the deep white matter [8,17], accurately predicting the precise locations of NET2 lesions for an individual patient is challenging. Nonetheless, in many clinical contexts, the goal is not solely to pinpoint exact locations but rather to provide clinicians with a range of plausible future images that indicate the general locations and counts of lesions under different treatments. Having this capability would represent an advance in personalized medicine and enhance clinical decision support.

To address these challenges, we turn to generative AI, particularly variational diffusion models, which have emerged as a powerful tool for medical image synthesis. Pioneering works have shown that diffusion models can generate high-fidelity medical images via progressive denoising [6,19,25], and subsequent refinements have enabled 3D diffusion models to preserve fine anatomical details and subtle variations [15,24]. However, despite several recent advances in generating high-quality brain MR images and disease progression over time [7,14,20], there are currently no generative architectures that (i) predict the appearance of new or enlarging pathological structures at future time points, (ii) perform causal inference on the resulting focal pathology maps based on real treatments, and (iii) efficiently train within a high-resolution voxel space.

In this work, we present the first treatment-aware diffusion framework that uses ControlNet to predict the evolution of NET2 lesion masks in MS. Our approach generates voxel-level predictions by restructuring 3D MRI data into pseudo-2D slabs, enabling a simple training framework free of the reliance on a separately trained VAE for image compression [19,24]. By integrating baseline imaging modalities (e.g., FLAIR, T2, and gadolinium-enhanced sequences) with treatment assignment, our model enables direct comparisons across treatment arms, while supporting multiple predictions at inference to capture variability in new disease activity.

Extensive experiments on a multi-centre dataset of 2131 patient 3D MRIs and corresponding lesion masks from randomized clinical trials for relapsing-remitting MS demonstrate that our model accurately predicts NET2 lesion masks for patients across treatments of six different efficacies. Moreover, we show that our model has the potential for real-world downstream tasks such as future lesion count and location estimation, binary lesion activity classification, and qualitative counterfactual generation. Our findings highlight the potential of data driven approaches in personalized medicine, opening the door to future research on treatment-aware generative models and their integration into clinical workflows for neurological diseases such as MS.

2 Methodology

Variational diffusion models (VDMs) [16] approximate complex data distributions by modeling a two-part stochastic process: a forward noising process and a learned reverse denoising process. The forward process gradually perturbs data by adding Gaussian noise:

$$\boldsymbol{z}_t = \alpha_\lambda \boldsymbol{x} + \sigma_\lambda \boldsymbol{\epsilon}, \quad \boldsymbol{\epsilon} \sim \mathcal{N}(0, \mathrm{I}), \tag{1}$$

where $\boldsymbol{x}$ is the original data sample, $\boldsymbol{z}_t$ is the noisy latent at timestep t, and λ is a parameterization of the signal-to-noise ratio (SNR) that governs the trade-off between clean signal and added noise for a given $t \in [0, 1]$. The reverse process aims to reconstruct $\boldsymbol{x}$ by learning to predict and subtract away the added noise, typically through a neural network $\hat{\boldsymbol{x}}_\theta(\boldsymbol{z}_t; t)$. The training objective minimizes a weighted mean squared error between the true data and the predicted data:

$$\mathcal{L}_{VDM} = \mathbb{E}_{\epsilon \sim \mathcal{N}(0,\mathrm{I}), \lambda \sim p(\lambda)} \left[\frac{w(\lambda)}{p(\lambda)} ||\boldsymbol{x} - \hat{\boldsymbol{x}}_\theta(\boldsymbol{z}_t; t)||_2^2 \right], \tag{2}$$

where $p(\lambda)$ is a SNR sampling distribution and $w(\lambda)$ is a weighting function that balances learning across SNRs during training.

UNet architectures [21] are widely used as the backbone for modeling the reverse distribution in image-based diffusion networks. To incorporate additional conditioning into the UNet architecture, a ControlNet [26], creates a duplicate of the UNet contraction path to serve as an adapter within the network. This trainable copy is connected to the original blocks via zero convolutions, which are 1×1 convolution layers initialized with zero weights and bias. Over time, gradients flow through the zero convolution layers, gradually allowing the adapter learn how to incorporate image conditions into the output. During training, ControlNet follows the VDM objective with additional conditioning inputs:

$$\mathcal{L}_{VDM} = \mathbb{E}_{\epsilon \sim \mathcal{N}(0,\mathrm{I}), \lambda \sim p(\lambda)} \left[\frac{w(\lambda)}{p(\lambda)} ||\boldsymbol{x} - \hat{\boldsymbol{x}}_\theta(\boldsymbol{z}_t, \boldsymbol{c}_e, \boldsymbol{c}_f; t)||_2^2 \right], \tag{3}$$

where $\boldsymbol{c}_e$ and $\boldsymbol{c}_f$ correspond to conditioning embeddings and images, respectively. To encourage the model to learn useful information from both conditioning styles, $\boldsymbol{c}_e$ and $\boldsymbol{c}_f$ are randomly masked during training with independent probabilities, forcing the network to attend to both sources of context.

The proposed method makes use of a voxel-level diffusion model that is pre-trained to learn the distribution of the NET2 labels, in combination with a ControlNet, which is conditioned on patient image(s) and associated labels (e.g. treatment), to predict a future, patient-specific NET2 mask. Both models are also conditioned on the treatment arm, which enables treatment specific predictions at inference. The overall architecture is depicted in Fig. 1, and described in more detail below.

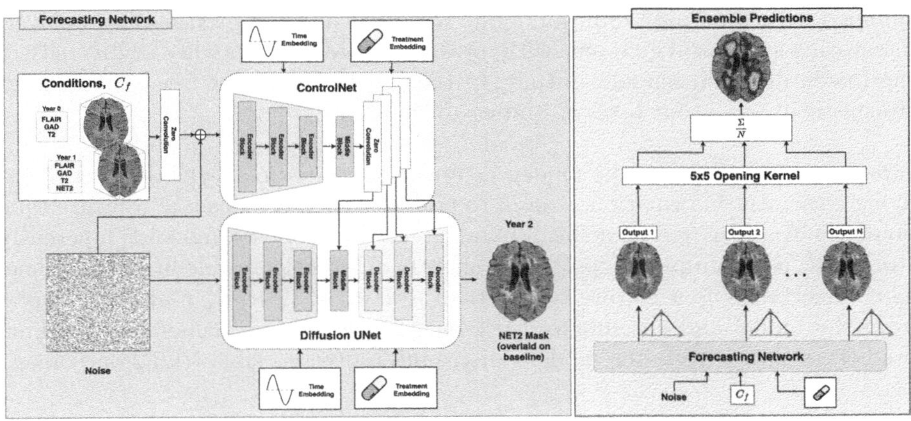

Fig. 1. Left: Model Architecture. A treatment-conditioned diffusion UNet is first pretrained to model the distribution of future NET2. A ControlNet is then trained to enable semantic conditioning based on MRI input. **Right: Inference Pipeline.** Multiple stochastic samples are ensembled to generate a treatment-aware NET2 mask prediction for a given patient.

Stage 1 - Pre-training the Diffusion Model: We first train a treatment-conditioned diffusion model to learn the distribution of future NET2 on each treatment arm. We combine the slice and channel dimensions of 3D MRI tensors ([b, c, s, h, w] $\rightarrow$ [b, c $\times$ s, h, w]), allowing the generation of 3D MRI and corresponding lesion labels with a 2D UNet [21]. Convolutional layers capture spatial relationships in the sagittal-coronal plane, while self-attention layers ensure consistency in the axial dimension. As modern accelerators offset the cost of scaling the number of channels [13], we are able to dramatically reduce the computational burden of storing 3D feature maps while allowing the use of a much higher capacity model at an equivalent batch size.

Conditioning diffusion models on text or categorical inputs can be addressed via architectural design, such that the denoising model becomes $\hat{\boldsymbol{x}}_\theta(\boldsymbol{z}_t, \boldsymbol{c}_e; t)$ where $\boldsymbol{c}_e$ is a categorically embedded patient treatment arm assignment. Following [2], we sum the class embeddings with the time embeddings that are injected into the ResBlocks of the UNet backbone. In order to be able to use classifier-free guidance at inference [12], we simultaneously train the diffusion model to learn the unconditional NET2 distribution by randomly dropping out the treatment (using a "null" embedding) 10% of the time. Moreover, we derive $p(\lambda)$ from an interpolated-shifted cosine schedule [13] and use a min-SNR [10] $w(\lambda)$ function.

Stage 2 - Training the ControlNet for Predicting Future NET2: We next train the ControlNet in order to enable semantic control over generation by fine-tuning a frozen copy of a pre-trained diffusion model encoder to process conditioning images, $\boldsymbol{c}_f$, in addition to class embeddings, $\boldsymbol{c}_e$, during denoising steps, such that the model becomes $\hat{\boldsymbol{x}}_\theta(\boldsymbol{z}_t, \boldsymbol{c}_e, \boldsymbol{c}_f; t)$. Outputs of the ControlNet encoder are summed with the frozen diffusion model's feature maps at each resolution layer, integrating structural guidance without compromising the quality or diversity of the frozen diffusion model's output. In the spirit of classifier-free guidance, we stochastically drop out both $\boldsymbol{c}_e$ and $\boldsymbol{c}_f$ during fine-tuning.

Inference: At inference, the model, conditioned on patient MRI and accompanying treatment information, is used to predict the NET2 mask at a future time point via classifier-free guided sampling. Since the proposed model is inherently stochastic, each sample can produce a slightly different plausible future outcome. This uncertainty is a feature rather than a flaw: by drawing multiple samples from the model and ensembling their outputs, we obtain a more robust final prediction that reflects the model's distribution over possible NET2 outcomes.

3 Dataset and Implementation Details

Dataset: Data are pooled from five randomized clinical trials that enrolled patients with relapsing-remitting MS (RRMS): BRAVO [23], OPERA 1/2 [11], and DEFINE [9]. Each patient sample consists of a Fluid Attenuated Inversion Recovery (FLAIR) image, lesion maps (T2 hyperintense and gadolinium-enhancing lesions), as well as relevant clinical features (treatment arm). NET2 lesion labels (identifying *new* T2 hyperintense lesions) between pre-treatment (w000) and week 48 (w048), as well as between w048 and week 96 (w096) are also available. We find that the vast majority of patients only have NET2 lesion activity occurring within a 15 slice slab taken from the center of the full volume. Thus, to reduce computational burden, we crop all MRI to dimensions of (15, 256, 256) in the axial, sagittal, and coronal axes, respectively.

In total, we construct a dataset consisting of 2131 samples belonging to the following trial arms: placebo (n = 327), laquinimod (no proven efficacy, NE, n = 237), avonex (low efficacy, LE, n = 271), interferon beta-1a subcutaneous (mild efficacy, MiE, n = 529), dimethyl fumarate (moderate efficacy, MoE, n = 193), and ocrelizumab (high efficacy, HE, n = 574). All MRI were acquired with $1 \times 1x3$ mm resolution at the following time points: pre-treatment (w000), one year (w048), and two years (w096). The data are split in an 80/10/10 ratio into train, validation, and test sets using stratification to preserve the treatment distribution across splits. Further statistics are provided in Table 1.

Implementation Details: For our UNet and ControlNet backbone, we use a similar encoder path as [2], with six down/up sampling stages and self-attention at 16^2 and 8^2 resolutions. The diffusion UNet is pre-trained for 100k iterations on FLAIR-NET2 image pairs from w096 conditioned on trial arm. This model is then frozen and used as the initialization for the ControlNet model, which we train for 40k iterations to predict w096 NET2 labels given w000 and w048 conditioning images and the trial arm. Given the small size of the NET2 and the relatively small number of patients that are active over time, we modify the loss function to up-weight NET2 lesion areas by a factor of 10, and secondly, we

Table 1. Samples, average NET2 lesion counts, and percentage of patients with NET2 lesion activity in the *training set* at 1 and 2 years post-treatment. w048 values are measured relative to pre-treatment; w096 values are relative to w048.

Treatment	Samples	Avg. NET2 Count		NET2 Activity (%)	
		w048	w096	w048	w096
Placebo	267	4.60	3.33	63.7	58.1
Laquinimod (NE)	184	5.51	4.49	64.7	57.1
Avonex (LE)	214	4.37	4.37	54.7	51.9
Interferon Beta-1a (MiE)	418	0.63	1.59	24.9	33.7
Dimethyl Fumarate (MoE)	153	0.60	1.03	19.6	26.1
Ocrelizumab (HE)	466	0.07	0.07	3.20	1.50
Total	1702				

train the ControlNet model on a balanced subset of the training data, ensuring an even distribution of model capacity across both active and inactive outcomes. Both models are trained on 4 A100 GPUs with a batch size of 16, and an Adam optimizer. In total, the models combine to contain 589M parameters.

4 Experiments and Results

We first analyze the spatial accuracy of the model's predicted NET2 lesion locations, followed by qualitative multi-inference results that highlight the model's stochastic yet trend-consistent behavior across treatment arms. Finally, we report the model's performance on downstream tasks that demonstrate its real-world utility. Given the lack of existing treatment-aware generative models, these experiments are evaluated against population-level statistical baselines to demonstrate relevant improvement.

4.1 Regional Spatial Accuracy

Accurate regional lesion localization is important for clinical interpretability. Although MS lesions play a critical role in disease assessment, they typically occupy less than 0.1% of total brain tissue and predominantly form within the cerebral white matter, and less prominently in the cortex [3,22]. To ensure that our proposed model generates plausible lesions in the correct regions without explicitly relying on disease-specific priors, we compare the predicted NET2 lesion regional locations against the ground truth regional locations for patients in the test set. This comparison permits accounting for stochastic nature of the exact location of new lesional formation. The regions considered are the cerebral white matter and cerebral cortex as segmented by SynthSeg [1].

For each brain region (cerebral white matter and cerebral cortex), we assign each patient a binary label: a ground truth label of 1 indicates the presence of at

Table 2. Comparison of the predictive regional accuracy of our method against a Monte Carlo baseline that uses positive class proportions from population-level statistics to predict the region that NET2 lesions will appear.

	BA ↑	Precision ↑	Recall ↑	F1 ↑
Cerebral-WM (Proposed Method)	**0.684**	**0.537**	**0.629**	**0.579**
Cerebral-WM (MC Baseline)	0.497	0.321	0.322	0.320
Cerebral-Cortex (Proposed Method)	**0.697**	**0.367**	**0.562**	**0.444**
Cerebral-Cortex (MC Baseline)	0.503	0.154	0.135	0.143

least one lesion in that region, while a label of 0 denotes its absence. Similarly, if the model predicts at least one lesion in a given region, the predicted label is set to 1, otherwise, it is set to 0. Table 2 compares the performance of the proposed model to a Monte Carlo baseline. For each region, the baseline predicts lesion presence by sampling from a Bernoulli distribution whose probability parameter is set to the empirical lesion prevalence for patients in the training set based on regions provided by SynthSeg segmentation [1]—approximately 33% for cerebral white matter and 10% for cerebral cortex. Our model outperforms this baseline, indicating its ability to predict the spatial distribution of NET2 lesions beyond what is achievable using population-level statistics alone. Note that no other baseline exists for this problem.

4.2 Exploring Ensembles of Multiple Inferences

To further demonstrate the model's ability to capture the stochastic nature of evolving MS disease activity, we perform inference several times for patients on different treatment arms. Figure 2 presents representative examples where, for two sample treatment conditions (placebo and a treatment with no proven efficacy), central 2D slices of the generated NET2 lesion masks are overlaid on baseline MRI scans. In both cases, the patients are active with NET2 appearing in the cerebral white matter around the ventricles. For the first patient, individual inferences from the model show predicted NET2 in this region, with some variability as to the precise location. For the second patient, an ensemble of 100 inferences shows a heatmap of possible future NET2 locations. In this case, there is overlap between an area of high probability (red) and the ground truth outcome, demonstrating clinical relevance in modeling the nature of the disease.

4.3 Relevant Downstream Tasks

To further evaluate our proposed model, we conduct two quantitative downstream tasks from the inferred NET2 labels. Note that the model was not explicitly trained for these tasks.

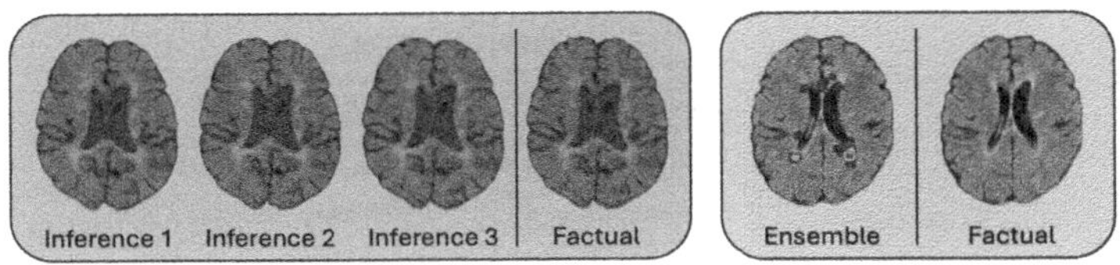

Fig. 2. Left: (Placebo) Three sample inferences compared to the factual outcome. Right: (NE) A heatmap of 100 inferences for a patient and the factual outcome.

Binary Activity Outcome Prediction: First, we assess the model's ability to predict future NET2 lesion activity by classifying whether a patient will have at least one NET2 lesion at w096, given pre-treatment and w048 data. To estimate the number of NET2 lesions in a mask generated by our model, we apply a morphological opening operation using a $1 \times 5 \times 5$ kernel to refine lesion boundaries, and then perform connected component analysis (CCA). For the ground truth count, we perform an identical CCA on the ground truth mask. Performance is measured using average precision (AP) and the area under the receiver operating characteristic curve (AUC). As a baseline, we include a Monte Carlo classifier that samples binary outcomes based on the observed frequency of positive cases at w096 in each treatment arm of the training set. Table 3, shows that our proposed model outperforms this baseline across each treatment group.

Table 3. Binary outcome prediction performance of the proposed model compared to a baseline built from population-level statistics. MC baseline metrics reported based on an average of 100 runs. Results reported on the test set

	Metric	Placebo	NE	LE	MiE	MoE	HE
Proposed Method	**AP ↑**	**0.783**	**0.735**	**0.623**	**0.647**	**0.436**	**0.083**
	AUC ↑	**0.826**	**0.662**	**0.536**	**0.705**	**0.607**	**0.868**
MC Baseline	**AP ↑**	0.513	0.577	0.517	0.361	0.342	0.019
	AUC ↑	0.495	0.505	0.487	0.507	0.489	0.509

NET2 Count Prediction: To further assess the models' performance, we compute the mean squared error (MSE) of the logarithm of the predicted NET2 lesion count, following [5]. Applying the logarithm helps reduce the influence of outliers, which is critical given the long-tailed distribution of NET2 lesion counts in MS. As a baseline, we use a predictor that assigns the average lesion count of each treatment group in the training set. As shown in Table 4, our model consistently outperforms this baseline across all treatment groups.

Table 4. MSE of log NET2 count against an average baseline on the test set.

	Placebo ↓	NE ↓	LE ↓	MiE ↓	MoE ↓	HE ↓
Proposed Method	**1.538**	**1.158**	**1.093**	**0.842**	**0.531**	**0.006**
Avg. Baseline	1.775	1.620	1.780	0.991	0.697	0.010

4.4 Exploring Counterfactual Predictions

Last, we explore the proposed model's potential to perform counterfactual image generation, specifically of a patient's future NET2 lesion mask, under different treatment conditions. We re-train the model using only pre-treatment images (w000) as conditioning information, and make predictions for a given patient under various treatments at w096. A few qualitative results are shown in Fig. 3. These results illustrate how treatments with different efficacies show reductions in NET2 lesions in the associated predicted images. While a detailed quantitative analysis of treatment effects is left for future work, these results highlight the promise of generative models in predicting future treatment-specific counterfactual medical images.

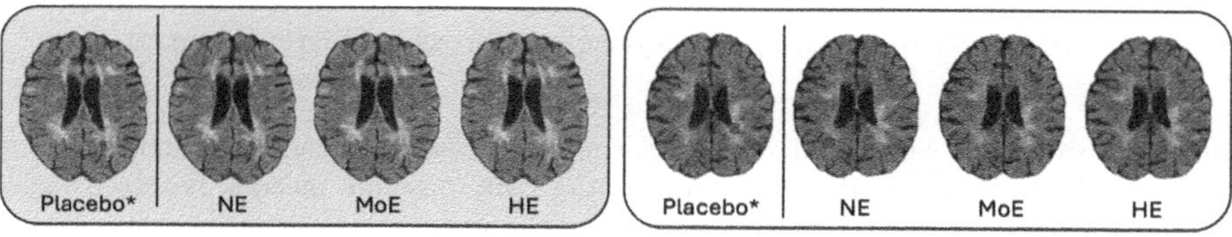

Fig. 3. Counterfactual predictions of future NET2 lesion masks for a single patient under treatments of varying efficacies, which increase from left to right. The model, conditioned only on pre-treatment data, reflects reduced future NET2 under more effective therapies, illustrating its potential for individualized treatment effect estimation.

5 Discussion

Predicting the future evolution of NET2 lesions in Multiple Sclerosis (MS) is a profoundly complex task, made challenging by both biological and technical factors. From a biological standpoint, the progression of MS is known to be highly heterogeneous [18]. Lesion development patterns vary significantly not only between population subgroups but also between individuals. Since NET2 lesion progression can follow multiple plausible trajectories, the observed outcome represents only one of several biologically reasonable possibilities. Consequently, standard metrics like accuracy, Dice score, or MSE may penalize valid predictions that deviate from this single outcome, overlooking clinically meaningful variation. This makes it difficult not only to model the disease but also to justify reasonable metrics for evaluating performance.

In addition to biological complexity, technical challenges inherent to medical imaging further complicate predictive modeling. MRI data, in particular, suffers from substantial variability due to differences in acquisition protocols, scanner hardware, spatial resolution, and contrast characteristics. While preprocessing steps like registration, normalization, and artifact correction are employed to mitigate these effects, such pipelines are imperfect and can introduce residual noise and inconsistencies that divert model capacity from true disease-related patterns. This is of course in addition to label noise and inter-rater variability in manual segmentation of NET2 lesions which can further impact the reliability of ground truth annotations.

It is also important to contextualize the Monte Carlo baselines used in our evaluations. While they may seem simplistic, they reflect real-world clinical heuristics: in the absence of individualized models, population-level statistics often form the basis for informing patients about prognosis. Thus, these baselines are not without value. However, as machine learning approaches mature, we anticipate that our models, baselines, and relevant metrics will evolve in sophistication and clinical relevance in personalized decision making.

Future research directions can, first, incorporate richer sources of patient-specific data as conditioning signals, including demographic, genetic, and clinical metrics for greater predictive accuracy. Second, a more formal, quantitative analysis of causal treatment effects at the individual level would extend the current qualitative findings presented in this work. Lastly, integrating MS-specific prognostic models, like [5], into the generation process could lead to improved performance by providing meaningful and structured prior knowledge [19].

6 Conclusion

In this work, we present the first treatment-aware conditional diffusion framework that forecasts the evolution of new and enlarging T2 (NET2) lesions in MS. Our approach employs a pseudo-2D representation of 3D MRI volumes, enabling high-resolution, voxel-level predictions without reliance on external anatomical priors. Comprehensive evaluations demonstrate that the model effectively predicts disease activity and provides interpretable, patient-specific insights into treatment responses. Our findings highlight the potential of generative AI to advance personalized medicine by allowing clinicians to visualize possible treatment responses, thereby moving towards AI-powered decision support for complex neurological disorders.

Acknowledgments. This investigation was supported by the International Progressive Multiple Sclerosis Alliance (PA-1412-02420) and the companies who generously provided the data: Biogen, Roche/Genentech, and Teva; the MS Society of Canada; the Natural Sciences and Engineering Research Council of Canada; Fonds de Recherche du Quebec: Nature et Technologies; the Canadian Institute for Advanced Research

(CIFAR) Artificial Intelligence Chairs program; Google Research; Calcul Quebec; the Digital Research Alliance of Canada; and Mila - Quebec AI Institute.

Disclosure of Interests. The authors have no competing interests to declare.

References

1. Billot, B., et al.: SynthSeg: segmentation of brain MRI scans of any contrast and resolution without retraining. Med. Image Anal. **86**, 102789 (2023). https://doi.org/10.1016/j.media.2023.102789
2. Dhariwal, P., Nichol, A.: Diffusion models beat GANs on image synthesis. Eprint arXiv:2105.05233 (2021)
3. Doyle, A., Elliott, C., Karimaghaloo, Z., Subbanna, N.K., Arnold, D.L., Arbel, T.: Lesion detection, segmentation and prediction in multiple sclerosis clinical trials. In: BrainLes@MICCAI (2017). https://api.semanticscholar.org/CorpusID:3440295
4. Durso-Finley, J., Falet, J.P., Mehta, R., Arnold, D.L., Pawlowski, N., Arbel, T.: Improving image-based precision medicine with uncertainty-aware causal models (2023). https://arxiv.org/abs/2305.03829
5. Durso-Finley, J., Falet, J.P., Nichyporuk, B., Douglas, A., Arbel, T.: Personalized prediction of future lesion activity and treatment effect in multiple sclerosis from baseline MRI. In: Konukoglu, E., Menze, B., Venkataraman, A., Baumgartner, C., Dou, Q., Albarqouni, S. (eds.) Proceedings of The 5th International Conference on Medical Imaging with Deep Learning. Proceedings of Machine Learning Research, vol. 172, pp. 387–406. PMLR (2022). https://proceedings.mlr.press/v172/durso-finley22a.html
6. Favero, G.M., Saremi, P., Kaczmarek, E., Nichyporuk, B., Arbel, T.: Conditional diffusion models are medical image classifiers that provide explainability and uncertainty for free (2025). https://arxiv.org/abs/2502.03687
7. Friedrich, P., Wolleb, J., Bieder, F., Durrer, A., Cattin, P.C.: WDM: 3D wavelet diffusion models for high-resolution medical image synthesis, pp. 11–21. Springer (2024). https://doi.org/10.1007/978-3-031-72744-3_2
8. Ge, T., Müller-Lenke, N., Bendfeldt, K., Nichols, T., Johnson, T.: Analysis of multiple sclerosis lesions via spatially varying coefficients. Ann. Appl. Stat. **8** (2014). https://doi.org/10.1214/14-AOAS718
9. Gold, R., et al.: Placebo-controlled phase 3 study of oral BG-12 for relapsing multiple sclerosis. N. Engl. J. Med. **367**(12), 1098–1107 (2012). https://doi.org/10.1056/NEJMoa1114287
10. Hang, T., et al.: Efficient diffusion training via min-snr weighting strategy. Eprint arXiv:2303.09556 (2024)
11. Hauser, S.L., et al.: Ocrelizumab versus interferon beta-1a in relapsing multiple sclerosis. N. Engl. J. Med. **376**(3), 221–234 (2017). https://doi.org/10.1056/NEJMoa1601277
12. Ho, J., Salimans, T.: Classifier-free diffusion guidance. Eprint arXiv:2207.12598 (2022)
13. Hoogeboom, E., Heek, J., Salimans, T.: Simple diffusion: end-to-end diffusion for high resolution images. Eprint arXiv:2301.11093 (2023)
14. Khader, F., Müller-Franzes, G., Tayebi Arasteh, S., et al.: Denoising diffusion probabilistic models for 3D medical image generation. Sci. Rep. **13**, 7303 (2023). https://doi.org/10.1038/s41598-023-34341-2

15. Kim, J., Park, H.: Adaptive latent diffusion model for 3D medical image to image translation: Multi-modal magnetic resonance imaging study. In: Proceedings of the IEEE/CVF Winter Conference on Applications of Computer Vision, pp. 7604–7613 (2024)
16. Kingma, D.P., Salimans, T., Poole, B., Ho, J.: Variational diffusion models. Eprint arXiv:2107.00630 (2023)
17. Lee, M.A., et al.: Spatial mapping of T2 and gadolinium-enhancing T1 lesion volumes in multiple sclerosis: evidence for distinct mechanisms of lesion genesis? Brain **122**(7), 1261–1270 (1999). https://doi.org/10.1093/brain/122.7.1261
18. Lucchinetti, C., Brück, W., Parisi, J., Scheithauer, B., Rodriguez, M., Lassmann, H.: Heterogeneity of multiple sclerosis lesions: implications for the pathogenesis of demyelination. Ann. Neurol. **47**(6), 707–717 (2000). https://doi.org/10.1002/1531-8249(200006)47:6<707::AID-ANA3>3.0.CO;2-Q
19. Puglisi, L., Alexander, D.C., Ravì, D.: Enhancing spatiotemporal disease progression models via latent diffusion and prior knowledge (2024). https://arxiv.org/abs/2405.03328
20. Rachmadi, M.F., Valdés-Hernández, M.C., Makin, S., et al.: Prediction of white matter hyperintensities evolution one-year post-stroke from a single-point brain MRI and stroke lesions information. Sci. Rep. **15**, 1208 (2025). https://doi.org/10.1038/s41598-024-83128-6
21. Ronneberger, O., Fischer, P., Brox, T.: U-net: convolutional networks for biomedical image segmentation. In: Navab, N., Hornegger, J., Wells, W.M., Frangi, A.F. (eds.) MICCAI 2015. LNCS, vol. 9351, pp. 234–241. Springer, Cham (2015). https://doi.org/10.1007/978-3-319-24574-4_28
22. Sepahvand, N.M., Arnold, D.L., Arbel, T.: CNN detection of new and enlarging multiple sclerosis lesions from longitudinal MRI using subtraction images. In: 2020 IEEE 17th International Symposium on Biomedical Imaging (ISBI), pp. 127–130 (2020). https://api.semanticscholar.org/CorpusID:218895124
23. Vollmer, T.L., et al.: A randomized placebo-controlled phase III trial of oral laquinimod for multiple sclerosis. J. Neurol. **261**(4), 773–783 (2014). https://doi.org/10.1007/s00415-014-7264-4
24. Wang, H., Liu, Z., Sun, K., Wang, X., Shen, D., Cui, Z.: 3D MedDiffusion: a 3D medical diffusion model for controllable and high-quality medical image generation. arXiv preprint arXiv:2412.13059 (2024)
25. Yoon, J.S., Zhang, C., Suk, H.I., Guo, J., Li, X.: SADM: sequence-aware diffusion model for longitudinal medical image generation. In: Frangi, A., de Bruijne, M., Wassermann, D., Navab, N. (eds.) IPMI 2023. LNCS, vol. 13939, pp. 388–400. Springer, Cham (2023). https://doi.org/10.1007/978-3-031-34048-2_30
26. Zhang, L., Rao, A., Agrawala, M.: Adding conditional control to text-to-image diffusion models. Eprint arXiv:2302.05543 (2023)

Mechanistic Learning with Guided Diffusion Models to Predict Spatio-Temporal Brain Tumor Growth

Daria Laslo[1,2(✉)], Efthymios Georgiou[3,7], Marius George Linguraru[4], Andreas M. Rauschecker[5], Sabine Müller[5,6], Catherine R. Jutzeler[1,2], and Sarah Brüningk[3,7]

[1] ETH Zurich, 8092 Zurich, Switzerland
[2] SIB Swiss Institute of Bioinformatics, 1015 Lausanne, Switzerland
daria.laslo@hest.ethz.ch
[3] Department of Radiation Oncology, Inselspital, Bern University Hospital and University of Bern, Bern, Switzerland
[4] Sheikh Zayed Institute for Pediatric Surgical Innovation, Children's National Medical Center, Washington, DC 20010, USA
[5] University of California San Francisco, San Francisco, CA 94143, USA
[6] University Children's Hospital Zurich, 8008 Zurich, Switzerland
[7] Department of Digital Medicine, University of Bern, Bern, Switzerland

Abstract. Predicting the spatio-temporal progression of brain tumors is essential for guiding clinical decisions in neuro-oncology. We propose a hybrid *mechanistic learning* framework that combines a mathematical tumor growth model with a guided denoising diffusion implicit model (DDIM) to synthesize anatomically feasible future MRIs from preceding scans. The mechanistic model, formulated as a system of ordinary differential equations, captures temporal tumor dynamics including radiotherapy effects and estimates future tumor burden. These estimates condition a gradient-guided DDIM, enabling image synthesis that aligns with both predicted growth and patient anatomy. We train our model on the BraTS adult and pediatric glioma datasets and evaluate on 60 axial slices of in-house longitudinal pediatric diffuse midline glioma (DMG) cases. Our framework generates realistic follow-up scans based on spatial similarity metrics. It also introduces *tumor growth probability maps*, which capture both clinically relevant extent and directionality of tumor growth as shown by 95th percentile Hausdorff Distance. The method enables biologically informed image generation in data-limited scenarios, offering generative-space-time predictions that account for mechanistic priors.

Keywords: Brain tumor modeling · Diffusion models · Mechanistic learning · Longitudinal MRI synthesis · Spatio-temporal prediction

1 Introduction

Longitudinal imaging is a cornerstone of the clinical workflow in neuro-oncology. Quantitative tumor segmentations enable disease burden and response classi-

B. Hou and T. S. Mathai (Eds.): LMID 2025, LNCS 16184, pp. 68–79, 2026.
https://doi.org/10.1007/978-3-032-16128-4_7

fication [3,22] in the longitudinal context. While mechanistic models of tumor volume dynamics have previously been explored [1,4,11,24], they strongly compress the complex spatial and anatomical aspects of radiographic data. Spatio-temporal predictions that capture both the extent of the tumor and its anatomical location are clinically more informative.

This is particularly critical for aggressive brain tumors located in sensitive regions, such as pediatric diffuse midline glioma (DMGs) [9]. Although generative modeling for brain tumor imaging has been conceptually established [17,18, 23], spatio-temporal generation of tumor growth remains underexplored, despite promising work in other neurological applications [12].

Denoising diffusion probabilistic models (DDPMs) [7], known for their high-fidelity image synthesis [2], allow conditioning and guidance via external inputs [8,10]. We leverage a guided denoising diffusion implicit model (DDIM) [20,23], directed by a regressor's gradient, to synthesize future scans reflecting increased tumor burden.

However, this straightforward approach lacks the temporal component, remaining conditioned on static targets (i.e., tumor burden). An additional challenge is the scarcity of extensive longitudinal data needed to train spatio-temporal generative models, particularly in rare and fatal diseases such as DMG.

In this work, we propose a mechanistic learning [14] approach that offers spatio-temporal brain tumor growth predictions given sparse and irregular temporal data. By integrating a mechanistic ordinary differential equation (ODEs) model of tumor dynamics with guided DDIMs, we are able to generate high fidelity multimodal MRI images. We evaluate image quality and prediction performance, demonstrating that our generated images conform with observed tumor growth patterns while preserving important anatomical structures. Our framework enables biologically informed generation of future tumor states, supporting anticipatory symptom management and therapy planning.

2 Methods

2.1 Guided Denoising Diffusion

DDPM's forward process is based on small amounts of noise added iteratively for L time steps to an input image x_0, obtaining a series of increasingly noisy images: $x_1, x_2, \ldots, x_L$. Generation is achieved by learning the reverse diffusion process, which iteratively denoises the corrupted image by obtaining x_{l-1} from x_l step by step until recovering x_0. Mathematically, this translates to learning the conditional distribution $p(x_{l-1}|x_l)$. A U-Net-based architecture (ϵ_θ) [19] is trained to approximate this distribution through a parametrized conditional distribution $p_\theta(x_{l-1}|x_l)$. Following the DDPM formulation [7,20], x_{l-1} can be generated from x_l via:

$$x_{l-1} = \sqrt{\bar{\alpha}_{l-1}} \left(\frac{x_l - \sqrt{1-\bar{\alpha}_l}\epsilon_\theta^{(l)}(x_l)}{\sqrt{\bar{\alpha}_l}} \right) + \sqrt{1-\bar{\alpha}_{l-1}-\sigma_l^2}\epsilon_\theta^{(l)}(x_l) + \sigma_l \epsilon_l \quad (1)$$

Setting $\sigma_l = 0$ results in a deterministic generative process, known as DDIM [20]. In our framework, we use the DDIM variant of our trained DDPM during inference to enable fast and reproducible image generation.

To synthesize axial brain slices with increased tumor burden, we follow the guidance approach from [2], training a separate regressor R to predict tumor size relative to brain area using images with varying noise levels (from 0 to L diffusion steps). Following [23], the regressor's gradient, scaled by a parameter s_R, directs the generation process toward images with the desired tumor size. The influence of this guidance is controlled by a weighting parameter s_R (regressor scale), which scales the gradient term, as described in the equation below.

$$\bar{\epsilon}_\theta^{(l)}(x_l) = \epsilon_\theta^{(l)}(x_l) - s_R\sqrt{1-\bar{\alpha}_l}\nabla_{x_l}R(x_l, l) \tag{2}$$

The regressor scale s_R can be modulated in two ways: dynamically based on the difference between the current regressor output and the target size, and statically via a scaling factor. Formally, we introduce $s_R = s_{R(ct)} * s_{R(dyn)}$, where $s_{R(ct)}$ is a constant scaling factor, and $s_{R(dyn)}$ is adaptively updated during sampling. As such, our framework enables the synthesis of brain MRIs of predefined tumor sizes, which we further combine with a mechanistic tumor growth model to simulate biologically plausible progression.

2.2 Mechanistic Model

We developed a mechanistic ODE model to characterize tumor growth dynamics in response to radiotherapy (RT). The model captures both intrinsic tumor growth and the delayed effects of RT-induced cell death, incorporating biologically motivated transitions in tumor composition [25].

Tumor Growth Prior to RT. Before the initiation of RT, tumor growth is assumed to follow exponential kinetics; a standard assumption for early-stage, untreated tumors, where resource limitations (e.g., oxygen, nutrients) have not yet induced growth saturation:

$$A(t) = A_0 e^{\lambda t}, \quad t < t_{\text{RTstart}} \tag{3}$$

where A_0 is the initial axial area and λ is the net growth rate.

Compartmental Partitioning at RT Onset. At the onset of RT ($t = t_{\text{RTstart}}$), the tumor is partitioned into two subpopulations to reflect differential cellular fates following irradiation: a surviving fraction (A_l) and a dying fraction (A_d):

$$A_l(t_{\text{RTstart}}) = S \cdot A(t_{\text{RTstart}}), \quad A_d(t_{\text{RTstart}}) = (1-S) \cdot A(t_{\text{RTstart}}) \tag{4}$$

The parameter $S \in [0, 1]$ represents the fraction of clonogenically surviving tumor and is conceptually analogous to surviving fraction metrics from radiobiology. This compartmentalization enables modeling of delayed cell death by mitotic catastrophe [6].

Post-RT Dynamics. Following RT, the surviving population continues to proliferate exponentially:

$$A_l(t) = A_l(t_{\text{RTstart}})e^{\lambda(t-t_{\text{RTstart}})} \tag{5}$$

A_d, representing cells that have sustained lethal DNA damage but may persist transiently, undergoes delayed shrinkage governed by a time-dependent decay rate:

$$A_d(t) = A_d(t_{\text{RTstart}})e^{\lambda'(t)(t-t_{\text{RTstart}})} \tag{6}$$

where $\lambda'(t) = \lambda_{\text{decay}} \tanh\left((t - t_{\text{RTstart}} - \delta) \cdot \text{slope}\right)$. The decay rate $\lambda'(t)$ captures the delayed onset and time course of RT-induced cell death. The tanh function enables a smooth transition from exponential growth to exponential decay at rate λ_{decay}, modulated by a delay parameter δ and a slope parameter controlling transition steepness. This form was chosen to reflect the gradual engagement of apoptotic and mitotic catastrophe pathways observed in irradiated tumors. The total tumor area for $t \geq t_{\text{RTstart}}$ is given by the sum of the two compartments $A(t) = A_l(t) + A_d(t)$.

This model structure enables physiologically interpretable parameters while capturing essential features of tumor response to RT, including subpopulation dynamics, delayed shrinkage, and regrowth potential.

Algorithm 1: Tumor Growth Modeling & Confidence Quantification

---

Input: A_0, λ, delay, slope, t, S (survival fraction)
Output: $A(t)$ (tumor area/volume)

Part 1: Compute Tumor Growth $A(t)$
if $t < t_{RTstart}$ **then**
  $A(t) \leftarrow A_0 \cdot e^{\lambda t}$
else
  $A_l(t_{\text{RTstart}}) \leftarrow S \cdot A(t_{\text{RTstart}})$
  $A_d(t_{\text{RTstart}}) \leftarrow (1 - S) \cdot A(t_{\text{RTstart}})$
  $\lambda_{\text{eff}} \leftarrow -\lambda \cdot \tanh\left(\frac{(t-t_{\text{RTstart}}-\text{delay})}{\text{slope}}\right)$
  $A_l(t) \leftarrow A_l(t_{\text{RTstart}}) \cdot e^{\lambda(t-t_{\text{RTstart}})}$
  $A_d(t) \leftarrow A_d(t_{\text{RTstart}}) \cdot e^{\lambda_{\text{eff}}(t-t_{\text{RTstart}})}$
  $A(t) \leftarrow A_l(t) + A_d(t)$
end
Part 2: Bootstrap Uncertainty Quantification
for $i \leftarrow 1$ **to** $N_{bootstrap}$ **do**
  $\tilde{A}_i \leftarrow A_{\text{observed}} + \mathcal{N}(0, \sigma^2_{\text{noise}})$
  $\theta_i \leftarrow \arg\min_\theta \sum_t \left|\tilde{A}_i(t) - A(t;\theta)\right|^2$
  $A_{\text{pred},i}(t) \leftarrow A(t;\theta_i)$
end
$A_{\text{median}}(t) \leftarrow \text{median}\{A_{\text{pred},i}(t)\}_{i=1}^{N_{\text{bootstrap}}}$
$A_{\text{CI}}(t) \leftarrow \text{percentile}\{A_{\text{pred},i}(t)\}_{i=1}^{N_{\text{bootstrap}}}$ at $[2.5\%, 97.5\%]$
return $A_{\text{median}}(t)$, $A_{\text{CI}}(t)$

2.3 Inference Using Mechanistic Learning

At inference time, we integrate the mechanistic tumor growth model into the guided diffusion framework to generate follow-up MRI scans. Given a series of tumor measurements, we fit a patient-specific model that allows temporal extrapolation of tumor burden. This model provides an estimate of the expected tumor size at the desired follow-up time.

To generate the follow-up scan, we start from the most recent reference image (e.g., from the last visit), which is noised by applying a fixed number of forward diffusion steps corresponding to a chosen noise level (NL). During the denoising (reverse diffusion) process, the extrapolated tumor size from the mechanistic model serves as a target for the regressor-based gradient guidance, allowing the model to synthesize an image consistent with both the input anatomy and the expected tumor progression. Specifically, we use the mechanistic model estimation to dynamically adjust the strength of the gradient ($\nabla_{x_l} R(x_l, l)$) based on the current evaluation compared to the set target through $s_{R(dyn)}$, that gets amplified by the constant scaling $s_{R(ct)}$ to define the final regressor scale s_R (Fig. 1).

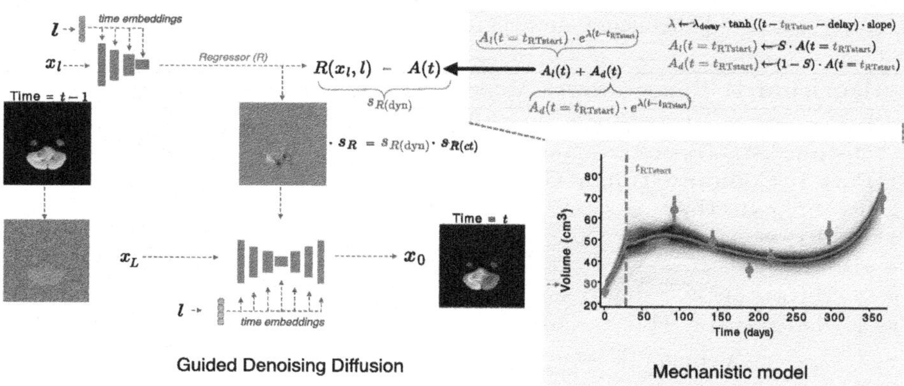

Fig. 1. Overview of the proposed method.

3 Experimental Results

3.1 Optimization of Guided Denoising Diffusion Model

Implementation Details. For training our guided diffusion framework, we used multiparametric MRI scans from 1125 adult and 105 pediatric high-grade glioma patients from BraTS 2023 Challenge [5,13]. All 2D axial slices containing brain tissue were included (n∼140,000). The DDPM was trained with 1,000 diffusion steps on the full dataset using a hybrid loss [15] over 175,000 training steps, with Adam optimizer. Independently, a regressor was trained on the same data to predict tumor size relative to brain volume, using 60,000 training steps, with Adam optimizer and mean squared error (MSE) loss. A validation set comprising 26 adult and 9 pediatric cases was used to compute validation loss and determine early stopping.

To evaluate inference performance, a set of longitudinally paired images (n = 185) from 29 pediatric DMG patients (before/during RT) collected from DMG Center Zurich was used. Ethical approval for this study was granted based on 2022-00312 by the North-West and Central Swiss Ethics Commission. Images were processed using the established BraTS pipeline [16], longitudinally coregistered, and manually reviewed based on automated tumor segmentation. For the optimization of *NL* and $s_{R(ct)}$, we set the target tumor size as known at the next imaging session. A grid search over a predefined parameter space was conducted.

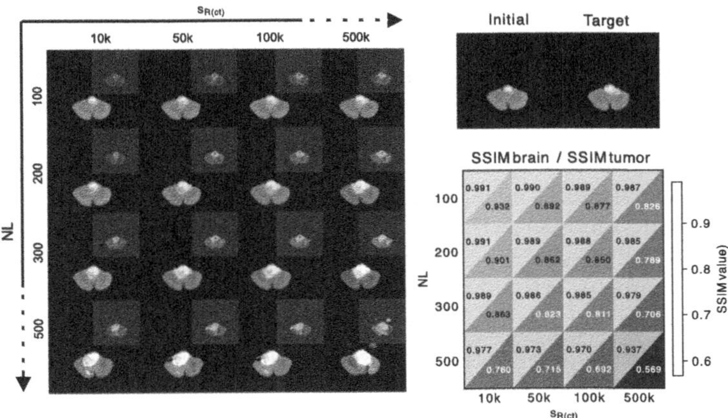

Fig. 2. Result for the grid search optimization of noise level (NL) and regressor scale ($s_{R(ct)}$). Generated images are shown along their difference maps to the starting image.

Evaluation. To assess the quality of generated images, we computed the Structural Similarity Index (SSIM) to the input image, separately within the tumor region and outside the extended tumor region. The regions were defined based on a dilated version (kernel adjusted to enable doubling of original tumor area) of the manual ground truth segmentation to account for the generated growth area. The region-specific evaluation was essential to capture focal changes (tumor) and the global consistency of the brain structure. Quantitative evaluation was complemented by qualitative visual assessments of the generated images compared to the actual follow-up scan, focusing on tumor progression patterns, anatomical feasibility, and the presence of artifacts or hallucinated structures.

Results. Figure 2 shows the generated images given a grid of NL and $s_{R(ct)}$ values, along corresponding difference maps comparing them to the input image (initial). Increasing NL leads to greater alterations in brain anatomy, with hallucinated structures appearing at the highest value ($NL = 500$). This is expected as a higher number of diffusion steps enables more flexibility in image synthesis. Similarly, increasing $s_{R(ct)}$ intensifies the tumor growth area, consistent with its radiological appearance on T2-weighted FLAIR, but excessive values ($\geq 100k$) produce unrealistic hyperintensities. The effects between the two parameters are cumulative, highlighted for extreme values ($s_{R(ct)} = 500k$ and $NL = 500$). The trends are consistent across the dataset as shown by the average region-specific SSIM values in Fig. 2. Based on both quantitative metrics and visual assessments of the generated images for the longitudinal pairs, we selected $NL = 200$ and $s_{R(ct)} = 50k$ for the downstream mechanistic learning predictions.

3.2 Mechanistic Modeling

Implementation Details. Our ODE model was fitted to longitudinal data from 8 pediatric DMG patients (60 slices) excluded from guided diffusion training, collected from DMG Center Zurich. Five equally spaced axial tumor slices, including the central one, were selected from the initial and largest tumor burden imaging sessions if a growth more than 10% was observed, totaling 60 2D axial slices treated independently. This patient and slice-specific approach captures individual tumor growth patterns and spatial heterogeneity. All implementations used `SciPy`'s [21] `odeint` for numerical solving across *pre* and *post* RT phases. Parameter estimation used the `lmfit` library[1] to minimize residuals between simulated and observed tumor trajectories. We performed 100 bootstrap iterations with 10% multiplicative Gaussian noise perturbation on tumor measurements. The model was re-fitted independently for each bootstrap replicate using different parameter constraints.

Evaluation. We evaluated our mechanistic model using two experimental setups: *all* used all available time-series data points to assess model accuracy in reproducing observed tumor growth patterns, while *train* used all points except the last, with the final time point reserved for testing predictive ability. The predictive horizon of our split is between [36, 141] days with a median of 67.

Performance was assessed using: 1) the coefficient of determination (R^2) to measure variance explained by our model, calculated between median predictions (over all bootstraps) and ground truth tumor area values and 2) normalized root mean square error (nRMSE) to quantify tumor size prediction error in the *train* setup, calculated between the median predicted value (over all bootstraps) and the ground truth at the last (unseen) timestep. RMSE was normalized by dividing by the ground truth reference tumor area value.

We selected the ODE fits with the best R^2 score on the *train* setup, as it represents the realistic clinical scenario where we have access to all measurements except the last and want to predict tumor area at the next time point.

[1] with nonlinear least-squares optimization.

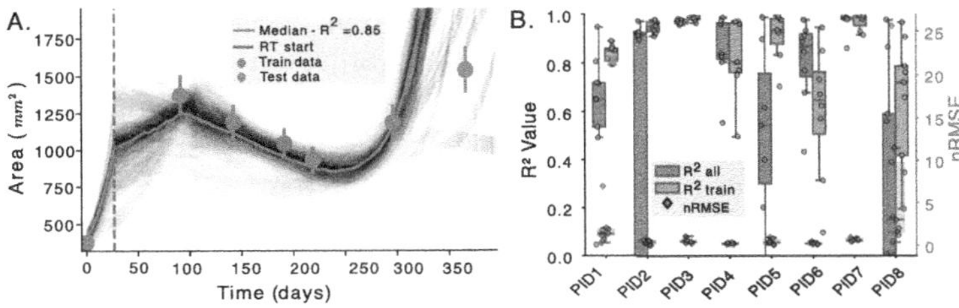

Fig. 3. A. Growth curve fitting (red) and estimation(gray). B. Mechanistic modeling performance metrics. Green cross highlights the example displayed in A. (Color figure online)

Results. Figure 3A presents a representative example of 100 fitted tumor area trajectories (bootstraps) for the *train* setup with the ground truth tumor area of the last time point (excluded from fitting) shown for reference. We observe that our ODE model solutions accurately capture tumor growth dynamics. Figure 3B summarizes the fitting and predictive performance of the ODE model. The *all* setup shows larger variance in R^2 values compared to the *train* setup. We focus on the *train* setup as it is clinically more relevant. The results demonstrate median R^2 values above 0.6 for the *train* setup, confirming that our ODE models effectively approximate tumor growth dynamics across patients and slices.

We further evaluate the predictive ability of these models through normalized RMSE (nRMSE) in Fig. 3B. The results show consistently low median nRMSE values (median over 60 slices is 0.468) with small variance across patients and slices, indicating reliable tumor area forecasts. This robust predictive performance, combined with good fitting ability, confirms that our ODE model effectively captures DMG growth patterns and provides an appropriate framework for modeling tumor dynamics in this data-sparse pediatric cancer.

3.3 Longitudinal Validation Using Mechanistic Learning

Implementation Details. We applied the integrated mechanistic learning framework to 60 2D axial brain slices from 8 additional pediatric patients (see *Mechanistic Modeling* subsection). Previous available tumor area measurements were used to parametrize the ODE model as described. The tumor sizes below 90th percentile of the bootstrap-estimated ($n = 100$) values were used as targets to simulate plausible tumor growth trajectories and their anatomical evolution. Tumor growth probability maps are computed as an average of difference maps between the generated and original tumor MRI scans that were previously binarized using the Otsu method. We refer to these as dynamic probability maps as they aggregate values from generations using varying target sizes. As a comparison, the framework was applied with the target set to the true tumor size. In this case, generations were repeated using randomly sampled noise for the forward diffusion process, enabling the aggregation of the resulting images as a static (same target size) tumor growth probability map.

Evaluation. We thresholded both the generated static (true tumor size) and dynamic (ODE-estimated tumor size) probability maps to generate binary masks. These were compared with the true target segmentation using the 95th percentile of the Hausdorff distance (HD95). For each slice, the HD95 was also computed between the initial and target image as a reference. A Wilcoxon-signed rank test was used to assess significance of HD95 change.

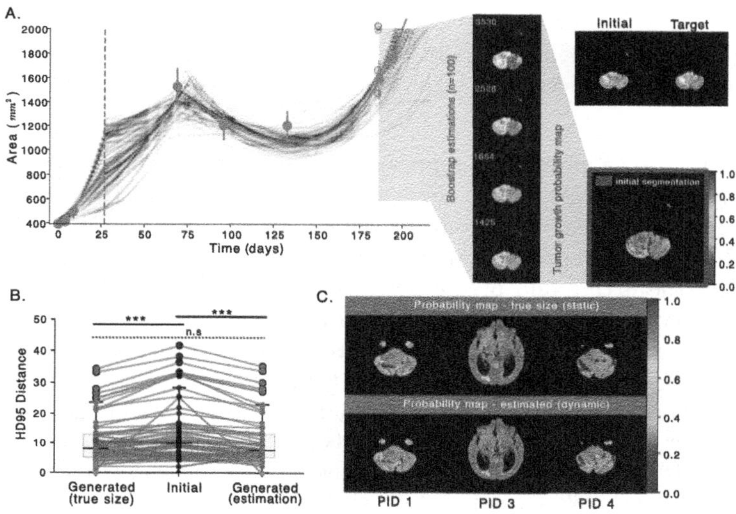

Fig. 4. A. Generating tumor growth probability maps based on bootstrap-estimated tumor sizes. B. Quantitative evaluation. C. Tumor probability map comparison. (Color figure online)

Results. Figure 4A highlights T2-FLAIR images generated towards having tumor sizes as sampled from the bootstrap distribution, illustrating varying degrees of tumor progression. This progression is mirrored in the aggregated probability maps, where high-probability regions (shown in red) correspond closely to the actual areas of tumor growth observed in the target images. Notably, the smooth spatial transitions across probability bands indicate consistent growth trajectories across tumor size variations, highlighting the robustness of the generative model. Quantitative evaluation in (Fig. 4B, C) supports that the HD95 distance between the generated mask and true targets is significantly lower ($p_{\text{value}} \leq 0.0005$) (green) than that between the initial image and the target. This supports the ability of the proposed framework to generate anatomically feasible tumor growth in the correct direction. Notably, the HD95 variability is higher in the case of thresholded static probability maps than when using the dynamic ones. This may reflect differences in tumor dynamics across samples, driven by variation in initial tumor sizes and the corresponding growth, which is factored in the dynamic maps through the bootstrap sampling covering a wider range of possible changes.

4 Conclusion

This work introduces a mechanistic learning framework for generative spatio-temporal tumor growth modeling, tailored towards data-sparse and unstructured scenarios. Our framework couples mechanistic ODE tumor growth models with generative models to construct a biologically informed image generation pipeline.

The mechanistic module, implemented via a biologically plausible ODE system, predicts the tumor area growth over time per patient and per slice. By feeding the mechanistic model's estimations to the diffusion model, we generate high fidelity brain tumor MRI scans conditioned on a target tumor area. Importantly, we aggregated generated images obtained from bootstrap estimates and synthesized both follow-up scans and tumor growth probability maps, which we showed to capture the directionality of tumor growth, making them highly valuable in the clinic.

Despite these promising results, some limitations remain. The mechanistic model assumes access to at least a minimal number of imaging time points to accurately capture tumor growth trajectories, while the guided diffusion network relies on using multiparametric MRI scans as input. Generalization could be improved through the application of the proposed methodology to single contrast which should be assessed via an ablation study. Moreover, future work could explore patient-specific fine-tuning of the guided diffusion process, including generating tumors of varying sizes and optimizing structural similarity (SSIM) between real and synthesized brain regions and tumors.

Taken together, our experiments evaluated each module of the framework separately and in combination, across different clinical setups, demonstrating the feasibility of such a hybrid approach. Our synergistic mechanistic learning framework shows promising results which pave a promising direction for biologically informed longitudinal tumor growth image generation. Our experiments illustrate the direct application of our framework in pediatric DMG patients.

Acknowledgments. This study was funded by Olga Mayenfisch Foundation, SNF (CRSK-3_229103, TMSGI3_225913), Chad Though Defeat Cancer Foundation.

Disclosure of Interests. The authors have no competing interests to declare that are relevant to the content of this article.

Code availability. The source code is available at: https://github.com/CAIROLab-Bern/Mechanistic-Learning-Brain-Tumor-Growth.

References

1. Brüningk, S.C., et al.: Intermittent radiotherapy as alternative treatment for recurrent high grade glioma: a modeling study based on longitudinal tumor measurements. Sci. Rep. **11**(1), 20219 (2021). https://doi.org/10.1038/s41598-021-99507-2
2. Dhariwal, P., Nichol, A.: Diffusion models beat GANs on image synthesis. Adv. Neural. Inf. Process. Syst. **34**, 8780–8794 (2021)

3. Erker, C., et al.: Response assessment in paediatric high-grade glioma: recommendations from the response assessment in pediatric neuro-oncology (RAPNO) working group. Lancet Oncol. **21**(6), e317–e329 (2020). https://doi.org/10.1016/S1470-2045(20)30173-X
4. Ezhov, I., et al.: Neural parameters estimation for brain tumor growth modeling. arXiv [q-bio.QM] (2019)
5. Fathi Kazerooni, A., et al.: BraTS-PEDs: results of the multi-consortium international pediatric brain tumor segmentation challenge 2023. J. Mach. Learn. Biomed. Imaging **3**(June 2025), 72–87 (2025). https://doi.org/10.59275/j.melba.2025-f6fg
6. Hazout, S., Oehler, C., Zwahlen, D.R., Taussky, D.: Historical view of the effects of radiation on cancer cells. Oncol. Rev. **19**, 1527742 (2025). https://doi.org/10.3389/or.2025.1527742
7. Ho, J., Jain, A., Abbeel, P.: Denoising diffusion probabilistic models. Adv. Neural. Inf. Process. Syst. **33**, 6840–6851 (2020)
8. Ho, J., Salimans, T.: Classifier-free diffusion guidance. arXiv [cs.LG] (2022)
9. Khalighi, S., Reddy, K., Midya, A., Pandav, K.B., Madabhushi, A., Abedalthagafi, M.: Artificial intelligence in neuro-oncology: advances and challenges in brain tumor diagnosis, prognosis, and precision treatment. NPJ Precis. Oncol. **8**(1), 80 (2024). https://doi.org/10.1038/s41698-024-00575-0
10. Konz, N., Chen, Y., Dong, H., Mazurowski, M.A.: Anatomically-controllable medical image generation with segmentation-guided diffusion models. arXiv [eess.IV] (2024)
11. Lipkova, J., et al.: Personalized radiotherapy design for glioblastoma: integrating mathematical tumor models, multimodal scans, and Bayesian inference. IEEE Trans. Med. Imaging **38**(8), 1875–1884 (2019). https://doi.org/10.1109/tmi.2019.2902044
12. Litrico, M., Guarnera, F., Giuffirda, V., Ravì, D., Battiato, S.: TADM: temporally-aware diffusion model for neurodegenerative progression on brain MRI. arXiv [eess.IV] (2024)
13. Menze, B.H., et al.: The multimodal brain tumor image segmentation benchmark (BRATS). IEEE Trans. Med. Imaging **34**(10), 1993–2024 (2015). https://doi.org/10.1109/TMI.2014.2377694
14. Metzcar, J., Jutzeler, C.R., Macklin, P., Köhn-Luque, A., Brüningk, S.C.: A review of mechanistic learning in mathematical oncology. Front. Immunol. **15**, 1363144 (2024). https://doi.org/10.3389/fimmu.2024.1363144
15. Nichol, A., Dhariwal, P.: Improved denoising diffusion probabilistic models (2021)
16. Pati, S., et al.: The cancer imaging phenomics toolkit (CaPTk): technical overview. Brainlesion **11993**, 380–394 (2020). https://doi.org/10.1007/978-3-030-46643-5_38
17. Peng, W., Adeli, E., Zhao, Q., Pohl, K.M.: Generating realistic 3D brain MRIs using a conditional diffusion probabilistic model (2022)
18. Pinaya, W.H.L., et al.: Brain imaging generation with latent diffusion models. arXiv [eess.IV] (2022)
19. Ronneberger, O., Fischer, P., Brox, T.: U-net: convolutional networks for biomedical image segmentation. In: Navab, N., Hornegger, J., Wells, W.M., Frangi, A.F. (eds.) MICCAI 2015. LNCS, vol. 9351, pp. 234–241. Springer, Cham (2015). https://doi.org/10.1007/978-3-319-24574-4_28
20. Song, J., Meng, C., Ermon, S.: Denoising diffusion implicit models (2020)
21. Virtanen, P., et al.: SciPy 1.0: fundamental algorithms for scientific computing in python. Nat. Methods **17**, 261–272 (2020). https://doi.org/10.1038/s41592-019-0686-2

22. Wen, P.Y., et al.: Updated response assessment criteria for high-grade gliomas: response assessment in neuro-oncology working group. J. Clin. Oncol. **28**(11), 1963–1972 (2010). https://doi.org/10.1200/JCO.2009.26.3541
23. Wolleb, J., Sandkühler, R., Bieder, F., Cattin, P.C.: The swiss army knife for image-to-image translation: multi-task diffusion models (2022)
24. Yankeelov, T.E., et al.: Clinically relevant modeling of tumor growth and treatment response. Sci. Transl. Med. **5**(187), 187ps9 (2013). https://doi.org/10.1126/scitranslmed.3005686
25. Zheng, D., et al.: Mathematical modeling in radiotherapy for cancer: a comprehensive narrative review. Radiat. Oncol. **20**(1), 49 (2025). https://doi.org/10.1186/s13014-025-02626-7

Longitudinal Brain Segmentation with Temporal Consistency for Neurodegenerative Analysis

Sheikh Adilina[1,2](✉), Leo Lebrat[1], Katie McMahon[1], Clinton Fookes[1], and Pierrick Bourgeat[2]

[1] Queensland University of Technology, Brisbane, QLD, Australia
[2] Australian e-Health Research Centre, CSIRO, Brisbane, QLD, Australia
adi010@csiro.au

Abstract. Magnetic resonance imaging (MRI) can detect subtle structural brain changes associated with neurodegenerative disorders up to a decade before clinical symptoms appear, making it a vital tool for early intervention. Longitudinal segmentation enables precise quantification of these changes; however, the task is challenging due to the faint nature of brain atrophy (0.5–1.0% per year) and the potential for inter-scan noise to obscure clinically meaningful patterns. In this work, we introduce two novel temporal consistency loss functions for longitudinal brain segmentation: a volume-based loss (VL) that promotes consistent volume trajectories without population-level normalization and a signed distance field (SDF) loss that enforces expected structural shrinkage without relying on age or time intervals. Unlike prior methods, our framework handles variable gaps between timepoints, making it more applicable to real-world clinical data. Our method is evaluated using group separation (Cohen's d), both cross-sectionally and longitudinally, which more accurately reflect longitudinal trends and provides a more clinically meaningful metric compared to traditional Dice scores. This provides a robust and adaptable framework for clinically meaningful longitudinal brain segmentation.

Keywords: Semi supervised learning · Longitudinal MR Imaging · Medical Image Segmentation

1 Introduction

Neurodegenerative disorders like Alzheimer's disease (AD) and Parkinson's disease (PD) are becoming more common with age and are typically marked by structural changes in the brain or atrophy [5,10,18]. These conditions are progressive and irreversible, and since there is currently no cure, early detection is crucial for treatments that can slow down their progression. One widely used non-invasive technique for investigating such disorders is magnetic resonance imaging (MRI). MRI can detect structural brain alterations up to 10 years before

B. Hou and T. S. Mathai (Eds.): LMID 2025, LNCS 16184, pp. 80–90, 2026.
https://doi.org/10.1007/978-3-032-16128-4_8

clinical symptoms appear, allowing early intervention with treatments such as anti-amyloid therapy [16].

While a single MRI scan already provides useful insights, using longitudinal data (multiple scans from the same person over time) facilitates the identification of subtle structural changes. This is particularly important for tracking change in brain volume over time through segmentation. In healthy aging, the brain undergoes an annual volumetric change of approximately 0.5–1.0%, and only specific regions (e.g. hippocampus) exhibit early signs of atrophy in conditions like mild cognitive impairment (MCI) and AD. Therefore, precise quantification of volumetric changes is crucial.

However, one of the major challenges in longitudinal analysis is the variability and noise present in MRI volumes, which can hide the underlying structural changes that need to be studied [11]. Moreover, the lack of annotated longitudinal datasets makes it harder to develop and test reliable deep learning based segmentation models.

Traditional segmentation tools like ANTs [17] and FreeSurfer [4] perform longitudinal segmentation using atlas-based preprocessing and subject-specific templates to ensure temporal consistency across time points. But these tools rely heavily on complex image registration and transformation steps, which can be computationally expensive and slow, particularly in multi-timepoint longitudinal studies. They also tend to perform poorly on noisy or low-resolution images, limiting their scalability to large clinical datasets.

Birenbaum et al. [2] were among the first to use longitudinal scans for MS lesion segmentation. Their method uses a baseline and a follow-up scan, aligns the scans using the Montreal Neurological Institute (MNI) template, and extracts candidate patches based on MR intensity and prior white matter maps. These patches are then processed through a CNN to estimate likelihood of lesion in each voxel. Denner et al. [3] built on this by introducing a deformable registration step after preprocessing. Since the images are already co-registered initially, this extra step helps capture significant structural changes like MS lesion progression, leading to better segmentation results. Some other models [8,15] go a step further by using fully automated convolutional networks for both registration and segmentation, also using two scans as input. Similarly, the JTRS framework [1] performs joint registration and segmentation, and Segis-Net [9] does the same in a single step. However, these methods remain limited to analyzing only two time points.

More recently, LongiSeg [13] was introduced to address the limitations of existing longitudinal segmentation methods by using a dataset with more than two time points. Built as an extension of nnU-Net [6], which still remains the state-of-the-art for cross-sectional segmentation, LongiSeg introduces diffusion-weighted blocks at each skip connection of the U-Net architecture [14]. These blocks focus on capturing temporal differences between consecutive scans, promoting temporally consistent segmentations. However, the method has some limitations. Firstly, it requires segmentation labels for all time points during training, which is often not feasible due to the limited availability of fully annotated longitudinal datasets. Secondly, it relies on access to previous scans even during inference, and cannot generate predictions from just one time point.

Another model, CFSegNet [19], takes a semi-supervised approach that makes use of unlabeled longitudinal data with multiple time points. It offers an advantage over LongiSeg by allowing inference from a single scan without requiring access to preceding or subsequent time points. The model incorporates volume-aware loss functions to guide the network in capturing anatomical changes over time, thereby enhancing temporal consistency. Nevertheless, CFSegNet assumes equidistant time intervals between scans and lacks the ability to handle variable gaps between follow ups. Moreover, the volume-aware losses employed do not explicitly account for inter-subject variability in intracranial volume (ICV), which can potentially limit their reliability in clinical settings.

Addressing the limitations mentioned above, our main contributions are as follows.

1. **Support for Irregular Longitudinal Sampling:** Our framework accommodates variable time gaps between visits, making it more applicable to real-world longitudinal datasets.
2. **Volume Change Modeling without Intracranial Normalization:** We propose a novel approach that focuses on the rate of change in anatomical volumes over time, reducing reliance on ICV normalization. This eliminates the need for potentially unreliable population-based corrections and enables more accurate temporal analysis through closer adaptation to individual trajectories.
3. **Signed Distance Field (SDF)-Based Loss Function:** Since brain volume always decreases with time, we provide an SDF-based loss function that penalizes structural expansions across timepoints. In contrast to methods that depend on pre-defined atrophy trajectories, our method learns deformation dynamics directly from the data, enabling subject-specific, geometry-aware regularization that generalizes better across brain regions and disease stages.

2 Method

2.1 Dataset and Preprocessing

The dataset consists of 811 subjects from the Alzheimer's Disease Neuroimaging Initiative (ADNI) [7] imaged using T1-weighted MRI. Among the 811 subjects, 688 had follow-ups ranging from 2 to 13 visits. Demographic information was also available for this dataset, including age, diagnosis, and specific time points at which the scans were acquired. The average age of the participants is 74 years, with a male to female ratio of 3:2. The dataset includes three diagnostic categories – healthy controls, MCI, AD– with a fairly balanced distribution across these groups. The FreeSurfer [4] cross-sectional pipeline was used to obtain the segmentations of all subjects to be used as pseudo-ground truth. The left and right hippocampi were used as regions of interest (ROIs). All volumes were cropped to a size of $96 \times 96 \times 96$ around the ROI. Bias correction and Z-score normalization were applied as preprocessing steps. For subjects with multiple time points, all the time points were rigidly aligned to a within-subject template as described in [12] to minimize bias.

2.2 Model Architecture

Inspired by the semi-supervised framework in CFSegNet [19], we trained our model using a two-stage strategy. The U-Net [14] architecture was used as the backbone for both stages. Similar to CFSegNet, the first stage was supervised using Dice loss, while the second stage was semi-supervised using Dice loss combined with four spatio-temporal constraints. More specifically, the Smoothness Constraint (SC) was used to guide the model in learning smooth changes in volume over time. The Age Constraint (AC), Volume Loss (VL), and Signed Distance Function (SDF) losses were used to enforce temporal volume reduction in anatomically meaningful ways. Each loss function is detailed below.

Smoothness Constraint (SC). This loss was originally used in CFSegNet [19] to promote smooth volume changes across time points. However, in the original paper, the equation was only applicable to time points that were equidistant from each other. We improved the equation so that it can now handle irregular time intervals, making it more applicable to real-world longitudinal datasets. The modified equation for SC loss is given in Eq. 1.

$$L_{\mathrm{SC}} = \mathrm{mean} \sum_{i=1}^{N-2} \sum_{\substack{j=2 \\ j>i}}^{N-1} \sum_{\substack{k=3 \\ k>j}}^{N} \left\| V^{t_j} - \frac{V^{t_i}|t_j - t_i| + V^{t_k}|t_k - t_j|}{|t_j - t_i| + |t_k - t_j|} \right\|^2, \tag{1}$$

where V^{t_i} denotes the volume of the ROI obtained at time point $t_i \in \mathbb{R}_+$.

Age Constraint (AC). The purpose of this loss function is to ensure that the model learns a decreasing volume pattern with age. In the CFSegNet [19] paper, the authors obtained the volume vs. age curve for each ROI using their dataset. Similarly, we computed the same curves for each of our ROIs from our dataset. The default cross-sectional nnU-Net was trained on the first time points of all subjects, which were previously labelled using FreeSurfer, and the model was subsequently applied to predict segmentations across the entire dataset, including unlabeled time points. The data points were plotted with absolute volume on the y-axis and age on the x-axis, as described in the original paper. The equations of the best-fit curves were computed for each ROI and used during training to guide the model toward age-consistent atrophy patterns.

Signed Distance Function (SDF) Loss. As atrophy can only lead to a volume reduction over time, the objective of this loss function is to promote structural shrinkage while penalizing any expansion. Given binary segmentation masks, $T_1, T_2 \in \{0,1\}^{D\times H\times W}$, at two consecutive time points, we compute their corresponding differentiable signed distance functions (SDFs), denoted as, $\phi_1 = \mathrm{SDF}(T_1)$ and $\phi_2 = \mathrm{SDF}(T_2)$.

The *shrinkage* $\mathcal{S}$ is defined for voxels that are present in T_1 but absent in T_2, and is defined as:

$$\mathcal{S}(T_1, T_2) = \sum_{\mathbf{x}} \mathbb{1}_{T_1(\mathbf{x})=1 \cap T_2(\mathbf{x})=0} \cdot \phi_2(\mathbf{x}), \tag{2}$$

where $\mathbb{1}_X$ is the indicator function of the set X

Conversely, the *expansion* $\mathcal{E}$ is defined for voxels that appear in T_2 but not in T_1:

$$\mathcal{E}(T_1, T_2) = \sum_{\mathbf{x}} \mathbb{1}_{T_1(\mathbf{x})=0 \cap T_2(\mathbf{x})=1} \cdot \phi_1(\mathbf{x}) \tag{3}$$

Together, these two functions are combined to define the SDF loss $\mathcal{L}_{\mathrm{SDF}}$; if expansion surpasses shrinkage, it implies structural growth:

$$\mathcal{L}_{\mathrm{SDF}} = \frac{1}{T-1} \sum_{t=2}^{T} \max\Big(\mathcal{E}(T_{t-1}, T_t) - \mathcal{S}(T_{t-1}, T_t),\, 0\Big) \tag{4}$$

Volume (VL) Loss. This loss aims to capture volume changes over time based on diagnosis-specific atrophy trends. Like done in the AC loss, the same nnU-Net predictions were used to estimate volume trajectories. However, rather than plotting volume versus age, we plotted the percentage volume change per year against mean volume for each diagnostic group (healthy controls, MCI and AD). Best-fit curves were computed for each group and used to guide the training process. The farther a prediction lies from its respective curve, the higher the loss, as shown in Fig. 1. The objective is to shift the predictions closer to the curve, ensuring they follow the expected atrophy patterns specific to each diagnosis. The loss function in Eq. 5 quantifies the average absolute deviation between the predicted change in volume ($\hat{y}_i$) and the corresponding expected change from the diagnosis-specific curve ($f(x_i)$), where x_i is the predicted mean volume.

$$\mathcal{L}_{\mathrm{VL}} = \frac{1}{N} \sum_{i=1}^{N} |\hat{y}_i - f(x_i)| \tag{5}$$

Unlike the AC loss, which relies on absolute volume and would require ICV normalization across the population for it to be reliable as individual brain volume differs, this approach eliminates the need for such normalization, making it more robust. Moreover, it also enables disease-specific modeling, which is not possible with the AC loss.

2.3 Implementation Details

The first stage (supervised) used only the first time point of each subject. In contrast, the second stage (semi-supervised) included only subjects with multiple

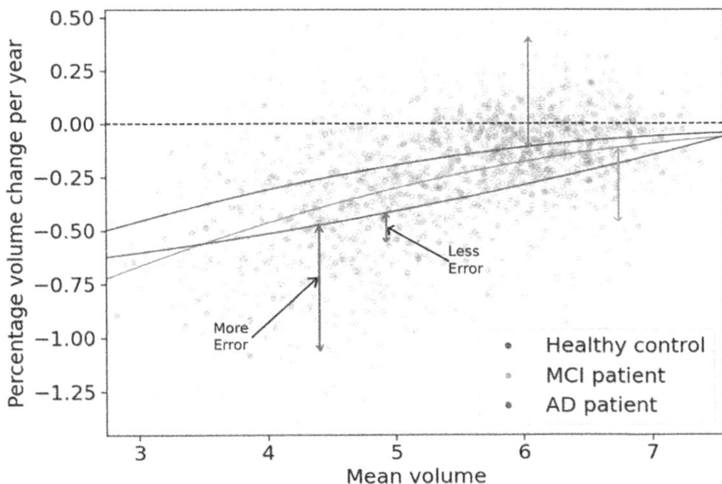

Fig. 1. Relationship between mean volume and yearly percentage volume change used to define the Volume Loss (VL) function.

time points. In Stage 1, the model was trained for up to 1,000 epochs using Dice loss, with pseudo ground truth generated by FreeSurfer [4]. In Stage 2, the model was refined for 150 epochs using Dice loss combined with the four spatiotemporal constraints described above. In this stage, no FreeSurfer labels were used; the Stage 1 model's predictions served as the pseudo ground truth.

Both stages used a batch size of 8 and were trained using the Adam optimizer with an initial learning rate of 0.001 and a weight decay of $1e^{-5}$. A ReduceLROnPlateau scheduler was applied to lower the learning rate by a factor of 0.8 if validation loss did not improve over three consecutive epochs. For both stages, the dataset was split in a 3:1:1 ratio for training, validation, and testing, respectively, and 5-fold cross-validation was used throughout.

2.4 Evaluation Metric

We use the Dice Similarity Coefficient (DSC) to evaluate segmentation performance on the labelled cross-sectional dataset at baseline. Dice scores are only reported for the first time point because we do not use any labelled data for the longitudinal dataset in Stage 2 of training. Since the longitudinal data is assumed to be unlabeled, we rely on alternative evaluation metrics that better capture temporal consistency and disease group separation. Specifically, we report Spearman's Rank Correlation Coefficient and Cohen's d. Spearman's rank correlation (ρ) is used to assess whether the predicted volumes of the ROIs decrease monotonically. It captures the overall trend in the data, independent of the scale of the predicted volumes. A higher ρ indicates that the model's predictions align with the excepted temporal pattern of shrinkage. Cohen's d measures how effectively the model can discriminate between diagnostic groups. We calculate it between pairs of groups: AD vs. MCI, AD vs. Healthy, and MCI vs. Healthy. We assess the separability of the rate of change of volume across diagnoses, as well as the absolute volume at the baseline time points. A higher value indicates better diagnostic group separation by the model.

3 Results and Discussion

We explored several temporal constraints as loss functions to improve the consistency of longitudinal segmentation. In particular, we improve the smoothness constraint (SC) loss originally used in CFSegNet [19]. In addition to improving the SC loss, we introduce two novel temporal regularization terms: a volume-based loss (VL), which encourages volume consistency over time, and a signed distance function (SDF) loss, which penalizes structural expansion at a pixel-level without requiring external information such as age or time gaps.

Table 1. Comparison of segmentation and temporal consistency metrics across models. Bold indicates best, underline second best.

Metric	FS	nnU-Net	LongiSeg	Baseline	CFSegNet	Ours
Dice (LH)	–	<u>86.426</u>	**86.554**	85.485	85.872	86.027
Dice (RH)	–	<u>87.073</u>	**87.175**	86.254	86.566	86.625
Spearman's rank	0.12	0.273	0.277	0.359	0.369	**0.387**
Rate of Change (Cohen's d)						
AD vs. MCI	0.118	0.239	0.255	0.217	**0.389**	<u>0.331</u>
AD vs. Healthy	0.252	0.628	0.643	0.857	<u>0.876</u>	**0.898**
MCI vs. Healthy	0.104	<u>0.329</u>	0.331	0.213	**0.359**	0.317
Volume of Baseline Timepoint (Cohen's d)						
AD vs. MCI	0.571	0.613	0.622	**0.63**	0.616	<u>0.625</u>
AD vs. Healthy	1.637	1.72	<u>1.728</u>	1.703	1.713	**1.739**
MCI vs. Healthy	0.947	0.956	<u>0.96</u>	0.944	0.956	**0.966**

A comparison of our proposed longitudinal framework (Ours) with state-of-the-art longitudinal models is shown in Table 1. It is important to note that we provided the reimplementation of the CFSegNet model, as the original paper did not provide any code. We changed the baseline model of CFSegNet and trained both CFSegNet and all our proposed models using the same U-Net backbone (Baseline) to ensure a fair architectural comparison. The CFSegNet results reported in Table 1 correspond to the version where the SC loss has been adapted to handle multiple time points. Additionally, pseudo labels needed to be obtained using FreeSurfer for all timepoints to train LongiSeg, as it requires the entire longitudinal dataset to be labeled, and no manually labeled dataset exists for the hippocampus. However, the Dice score for all methods, including LongiSeg,

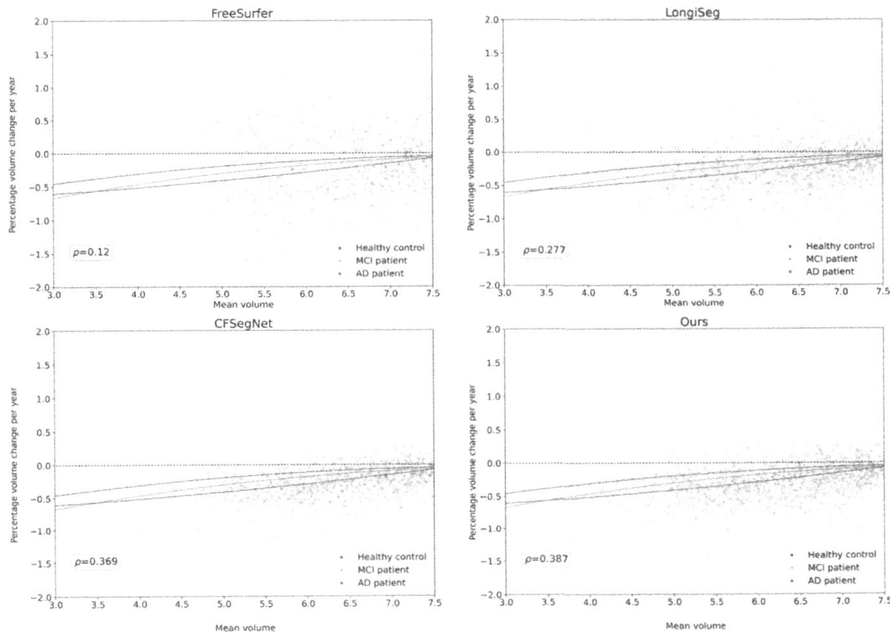

Fig. 2. Visual comparison of Spearman's rank correlation from FreeSurfer (top-left), LongiSeg (top-right), CFSegNet (bottom-left) and Ours (bottom-right).

was computed only at the baseline time point to ensure fairness, since for all the other models, only the labeled data at baseline was used for training, and the longitudinal training was done on unlabeled data. All other evaluation metrics, including Spearman's rank correlation and Cohen's d, were computed across all timepoints. Experimental results in multiple evaluation metrics demonstrate that, while LongiSeg achieves the highest Dice score, it performs poorly in clinically meaningful metrics, with poor group separability using baseline volumes and volumetric rate of change. This finding highlights a key insight: a high Dice score does not necessarily imply reliability or usefulness in a clinical context. A visual comparison of Spearman's rank correlation is provided in Fig. 2, illustrating that the better the model, the closer the predictions are to the curves.

Table 2 shows the experimental results of adding each loss function to our baseline U-Net model and illustrates how they contribute to improved temporal consistency. For simplicity, all loss function coefficients were set to 1 in our current implementation. Although combining all the loss functions in our final model results in strong performance, a more effective weighting of these losses could further improve the temporal consistency of the model and may ultimately produce a model that outperforms all other variants in all aspects. From Table 2, we can also observe that the AC loss performs very well in terms of volume separability at the baseline time point. While this effectively leverages the known age-volume

Table 2. Ablation study of model loss functions. The best-performing model is shown in bold. A red-to-green color gradient illustrates the relative performance of each model, with green indicating better results.

Metric	Baseline	SDF	VL	SDF+VL	SDF+VL +SC	SDF+VL +AC	SDF+VL +AC+SC
Dice (LH)	85.485	85.797	85.908	85.99	85.952	85.773	**86.027**
Dice (RH)	86.254	86.386	86.566	86.6	86.518	86.342	**86.625**
Spearman's rank	0.359	0.369	0.36	0.372	0.37	0.379	**0.387**
Rate of Change (Cohen's d)							
AD vs. MCI	0.217	0.317	0.311	0.323	0.3	0.302	**0.331**
AD vs. Healthy	0.857	0.872	0.875	0.856	0.862	0.892	**0.898**
MCI vs. Healthy	0.213	0.326	0.278	**0.341**	0.304	0.325	0.317
Volume of Baseline Timepoint (Cohen's d)							
AD vs. MCI	0.63	0.601	0.609	0.624	0.627	**0.64**	0.625
AD vs. Healthy	1.703	1.697	1.714	1.732	1.696	**1.766**	1.739
MCI vs. Healthy	0.944	0.955	0.958	0.963	0.945	0.962	**0.966**

relationship in hippocampal structures, this may not necessarily extend to other applications such as white matter lesions, where there is no direct association between lesion volume and age. In contrast, the VL and SDF loss functions we introduce are applicable in scenarios where no such age correlation exists. These losses rely solely on intrinsic volume and shape characteristics, enhancing their utility across diverse clinical contexts.

4 Conclusion

We presented a comprehensive framework for improving longitudinal brain segmentation using novel temporal loss functions. Our proposed volume-based and SDF-based loss functions don't rely on external factors like age, time intervals, or population-level normalization. This makes them more flexible and applicable to other brain structures and clinical tasks beyond hippocampal segmentation. In addition, by using Spearman's rank correlation and Cohen's d, we can better capture meaningful trends over time and distinguish between clinical groups—insights that align more closely with how diseases actually progress. Unlike metrics like Dice, which only measure overlap at a single time point, these longitudinal metrics reflect the true temporal behavior of the model. Altogether, our combination of generalizable loss functions and clinically relevant evaluation metrics provides a robust, adaptable framework for longitudinal brain segmentation across a variety of datasets and use cases.

Given these strengths, our method enables the investigation of lesions or anatomical regions in any longitudinal dataset where only partial ground truth is available. Future work will focus on applying this model to areas such as the entorhinal cortex, where anatomical boundaries are less clearly defined than in the hippocampus. In such challenging cases, our proposed volume-based and SDF-based constraints are expected to be even more beneficial.

Acknowledgments. This project is supported by a joint CSIRO–University PhD Scholarship between the Centre for Biomedical Technologies at Queensland University of Technology and the Australian e-Health Research Centre, CSIRO. Additional support was provided through a CSIRO Top-up Scholarship. We gratefully acknowledge the computational resources and expertise provided by CSIRO IMT Scientific Computing.

Code Availability. The implementation of our method is publicly available at: https://github.com/SkAdilina/MICCAI-2025.

Disclosure of Interests. The authors have no competing interests to declare that are relevant to the content of this article.

References

1. Barzilay, N., Nelkenbaum, I., Konen, E., Kiryati, N., Mayer, A.: Neural registration and segmentation of white matter tracts in multi-modal brain MRI. In: Karlinsky, L., Michaeli, T., Nishino, K. (eds) ECCV 2022 Workshops. LNCS, pp. 252–267. Springer, Cham (2023)
2. Birenbaum, A., Greenspan, H.: Multi-view longitudinal CNN for multiple sclerosis lesion segmentation. Eng. Appl. Artif. Intell. **65**, 111–118 (2017). https://doi.org/10.1016/j.engappai.2017.06.006
3. Denner, S., et al.: Spatio-temporal learning from longitudinal data for multiple sclerosis lesion segmentation (2020). arXiv preprint arXiv:2004.03675. https://doi.org/10.48550/arXiv.2004.03675
4. Fischl, B.: Freesurfer. Neuroimage **62**(2), 774–781 (2012). https://doi.org/10.1016/j.neuroimage.2012.01.021
5. Hou, Y., et al.: Ageing as a risk factor for neurodegenerative disease. Nat. Rev. Neurol. **15**, 565–581 (2019). https://doi.org/10.1038/s41582-019-0244-7
6. Isensee, F., Jaeger, P., Kohl, S., Petersen, J., Maier-Hein, K.: nnU-Net: a self-configuring method for deep learning-based biomedical image segmentation. Nat. Methods **18**(2), 203–211 (2021)
7. Jack, C.J., et al.: The alzheimer's disease neuroimaging initiative (ADNI): MRI methods. J. Magn. Reson. Imaging **27**(4), 685–691 (2008). https://doi.org/10.1002/jmri.21049
8. Krüger, J., et al.: Fully automated longitudinal segmentation of new or enlarged multiple sclerosis lesions using 3D convolutional neural networks. NeuroImage Clin. **28**, 102445 (2020). https://doi.org/10.1016/j.nicl.2020.102445
9. Li, B., et al.: Longitudinal diffusion MRI analysis using segis-net: a single-step deep-learning framework for simultaneous segmentation and registration. Neuroimage **235**, 118004 (2021). https://doi.org/10.1016/j.neuroimage.2021.118004
10. Li, G., et al.: The age-specific comorbidity burden of mild cognitive impairment: a us claims database study. Alzheimers Res. Ther. **15**, 211 (2023). https://doi.org/10.1186/s13195-023-01358-8

11. Liu, L., Wolterink, J., Brune, C., Veldhuis, R.: Anatomy-aided deep learning for medical image segmentation: a review. Phys. Med. Biol. **66**(11), 11TR01 (2021). https://doi.org/10.1088/1361-6560/abfbf4
12. Reuter, M., Schmansky, N., Rosas, H., Fischl, B.: Within-subject template estimation for unbiased longitudinal image analysis. Neuroimage **61**(4), 1402–1418 (2012). https://doi.org/10.1016/j.neuroimage.2012.02.084
13. Rokuss, M., et al.: Longitudinal segmentation of MS lesions via temporal difference weighting (2024). arXiv preprint arXiv:2409.13416
14. Ronneberger, O., Fischer, P., Brox, T.: U-net: Convolutional networks for biomedical image segmentation (2015). arXiv preprint arXiv:1505.04597
15. Salem, M., et al.: A fully convolutional neural network for new t2-w lesion detection in multiple sclerosis. NeuroImage Clin. **25**, 102109 (2020)
16. Tondelli, M., Wilcock, G., Nichelli, P., De Jager, C., Jenkinson, M., Zamboni, G.: Structural MRI changes detectable up to ten years before clinical alzheimer's disease. Neurobiol. Aging **33**(4), 825.e25-825.e36 (2012). https://doi.org/10.1016/j.neurobiolaging.2011.05.018
17. Tustison, N., et al.: The ants cortical thickness processing pipeline. In: Medical Imaging 2013: Biomedical Applications in Molecular, Structural, and Functional Imaging, vol. 8672, p. 86720K. SPIE, Bellingham (2013). https://doi.org/10.1117/12.2007128
18. Vinke, E., et al.: Trajectories of imaging markers in brain aging: the rotterdam study. Neurobiol. Aging **71**, 32–40 (2018). https://doi.org/10.1016/j.neurobiolaging.2018.07.001
19. Wei, J., Shi, F., Cui, Z., Pan, Y., Xia, Y., Shen, D.: Consistent segmentation of longitudinal brain MR images with spatio-temporal constrained networks. In: de Bruijne, M., Cattin, P.C., Cotin, S., Padoy, N., Speidel, S., Zheng, Y., Essert, C. (eds.) MICCAI 2021. LNCS, vol. 12901, pp. 89–98. Springer, Cham (2021). https://doi.org/10.1007/978-3-030-87193-2_9

Two-Stage Decoupling Framework for Variable-Length Glaucoma Prognosis

Yiran Song[1], Yikai Zhang[1], Silvia Orengo-Nania[1], Nian Wang[2], Fenglong Ma[3], Rui Zhang[1], Yifan Peng[4], and Mingquan Lin[1](✉)

[1] University of Minnesota, Minneapolis, MN, USA
{songyiran,zhan9191,sorengon,zhan1386,lin01231}@umn.edu

[2] UT Southwestern Medical Center, Dallas, TX, USA
nian.wang@utsouthwestern.edu

[3] The Pennsylvania State University, University Park, PA, USA
fenglong@psu.edu

[4] Weill Cornell Medicine, New York, NY, USA
yip4002@med.cornell.edu

Abstract. Glaucoma is one of the leading causes of irreversible blindness worldwide. Glaucoma prognosis is essential for identifying at-risk patients and enabling timely intervention to prevent blindness. Many existing approaches rely on historical sequential data but are constrained by fixed-length inputs, limiting their flexibility. Additionally, traditional glaucoma prognosis methods often employ end-to-end models, which struggle with the limited size of glaucoma datasets. To address these challenges, we propose a Two-Stage Decoupling Framework (TSDF) for variable-length glaucoma prognosis. In the first stage, we employ a feature representation module that leverages self-supervised learning to aggregate multiple glaucoma datasets for training, disregarding differences in their supervisory information. This approach enables datasets of varying sizes to learn better feature representations. In the second stage, we introduce a temporal aggregation module that incorporates an attention-based mechanism to process sequential inputs of varying lengths, ensuring flexible and efficient utilization of all available data. This design significantly enhances model performance while maintaining a compact parameter size. Extensive experiments on two benchmark glaucoma datasets—the Ocular Hypertension Treatment Study (OHTS) and the Glaucoma Real-world Appraisal Progression Ensemble (GRAPE), which differ significantly in scale and clinical settings, demonstrate the effectiveness and robustness of our approach. Code is available: MICCAI LMID Repository.

Keywords: Glaucoma Prognosis · Variable-length Sequences · Self-Supervised Learning · Attention-Based Temporal Aggregation

1 Introduction

Glaucoma is one of the leading causes of irreversible blindness worldwide [1]. Unlike glaucoma detection [2,4,16,18,25], which evaluates the current state of

B. Hou and T. S. Mathai (Eds.): LMID 2025, LNCS 16184, pp. 91–101, 2026.
https://doi.org/10.1007/978-3-032-16128-4_9

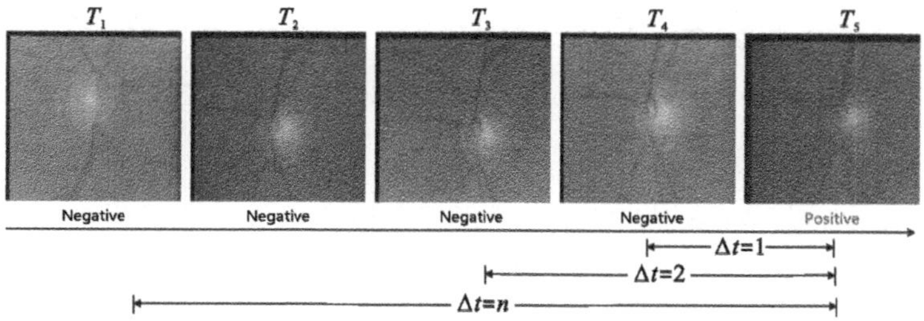

Fig. 1. Sequential fundus images of a patient of Δt

the disease based on existing data, glaucoma prognosis analyzes historical data to anticipate future disease progression. With the increasing global prevalence of glaucoma, accurate prognostic models are crucial to identify at-risk patients, enabling timely interventions, and advancing personalized treatments to prevent vision loss [11,15,22]. Several deep learning-based methods have been proposed for the prognosis of glaucoma [8,13,14,20,21,24]. For instance, Lin et al. [17] introduced a multi-scale multi-structure Siamese network (MMSNet) to estimate glaucoma progression using only the first and most recent fundus images. However, relying on only two images limits the ability to fully capture disease evolution over time. Li et al. [15] constructed a dataset comprising sequential fundus images and developed DeepGF, an LSTM-based model that learns spatial-temporal patterns from the fundus image sequence of a patient. DeepGF predicts the probability of progression of glaucoma in the next step, but does not accurately predict when that progression will occur. To improve temporal modeling, Hu et al. [11] proposed GLIM-Net, a Transformer-based network designed for irregularly sampled fundus image sequences, incorporating two time-related modules to regulate predictions.

Despite these advancements, two key challenges remain in glaucoma prediction: **1. The rigidity of fixed-length time series design significantly affects model flexibility.** Prognosis models with fixed-length designs struggle because glaucoma datasets exhibit a wide range of temporal spans across samples, making them highly sensitive to the chosen sequence length Δt (Fig. 1). A small Δt may result in insufficient accuracy while underutilizing longer-sequence data. Conversely, a larger Δt may cause some samples to be discarded, while excessive padding can degrade training performance. **2. The conflict between the large parameter scale of end-to-end models and the limited size of the glaucoma dataset.** End-to-end models require simultaneous learning of both feature extraction and prediction components. However, glaucoma datasets are typically small, and the need for longitudinal data to train sequence models further reduces the number of available patient-level samples. This constraint makes it difficult to effectively train models with large parameters, increasing the risk of severe overfitting.

To address these challenges, we propose a two-stage decoupling framework (TSDF) for glaucoma prognosis (Fig. 2) that separates feature representation

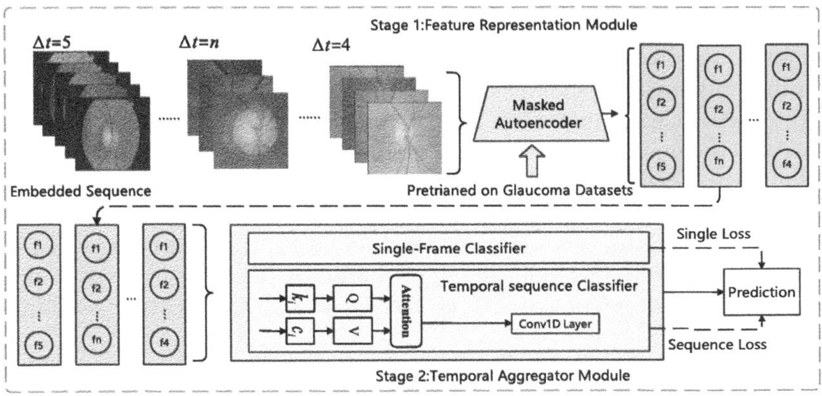

Fig. 2. The architecture of Our Two-Stage Decoupling Framework (TSDF).

from temporal aggregation. The representation learning module leverages self-supervised learning to fully utilize the available dataset and generate effective representations. Since self-supervised learning does not require labeled data, it enables training on multiple datasets of related diseases, enhancing feature quality for glaucoma prognosis. The temporal aggregator module incorporates an attention-based mechanism, efficiently aggregating variable-length time-series information while reducing model complexity and mitigating overfitting. Its dual-path design captures both single-frame and sequential-frame information. To the best of our knowledge, TSDF is the first framework capable of handling variable-length time-series inputs while simultaneously capturing both single-frame and time-series frame information for glaucoma prognosis. We evaluate TSDF on two publicly available datasets, the Ocular Hypertension Treatment Study (OHTS) [5,10] and the Glaucoma Real-world Appraisal Progression Ensemble (GRAPE) [12], demonstrating its effectiveness in handling variable-length sequences and achieving superior prediction performance. In contrast to traditional prediction models, our approach employs a decoupled training strategy that separates feature encoding and diagnosis. This approach enables small datasets to benefit from the extensive knowledge of large datasets during feature encoding, achieving superior patch encoding performance. Additionally, the diagnosis stage requires fewer training parameters, which reduces computational demands and enhances training efficiency, better aligning with practical application requirements.

2 Method

2.1 Self-supervised Masked Autoencoders for Feature Representation Learning

We propose leveraging self-supervised learning via a masked autoencoder (MAE) for feature extraction. Specifically, we employ the MAE [6] framework, a state-of-the-art self-supervised approach that enables the model to learn robust feature representations by reconstructing masked portions of the input data. The MAE randomly masks a significant fraction of the input patches. The encoder is trained to extract meaningful latent representations, while a lightweight decoder reconstructs the missing content. The training objective is to minimize the reconstruction error, encouraging the encoder to capture high-level semantic features. Once trained, the decoder is discarded, and the encoder is retained to generate feature representations. The datasets used for MAE training consist of patches densely sampled from a single image of temporal data. In each training iteration, a random subset of patches is masked to enhance feature learning and generalization.

The key advantage of this approach is that it does not require additional supervisory information, relying solely on the inherent structure of the data. This characteristic enables its application to glaucoma datasets with varying levels of supervision, allowing for the integration of multiple datasets for joint training. By aggregating multiple datasets for joint training, the model benefits from increased data diversity and sample size, thereby reducing the likelihood of overfitting typically observed with limited training data.

2.2 The Dual-Path Temporal Aggregator Module

Unlike previous approaches that separately learn single-frame classifiers or sequence classifiers, TSDF utilizes a dual-path architecture to **simultaneously optimize both the single-frame classifier and the sequence classifier**.

Let $T = \{I_0, ..., I_{M-1}\}$ represent a temporal sequence of frames. Given a feature representation function ϕ, each frame y_i is transformed into an embedding:

$$\mathbf{f}_i = \phi(I_i) \in \mathbb{R}^{D\times 1}, \quad i = 0, ..., M-1. \tag{1}$$

Single-Frame Classification Path: The first path applies a single-frame classifier to each frame embedding:

$$Output_{single}(T) = \mathbf{L}_0\mathbf{f}_0, ..., \mathbf{L}_{M-1}\mathbf{f}_{M-1} \tag{2}$$

where the weight matrix $\mathbf{L}_i$ is the instance classifier on each instance embedding.

Sequence Classification Path: It aggregates frame embeddings into a sequence representation. Each frame embedding $\mathbf{f}_i$ is transformed into two vectors:

$$\mathbf{k}_i = \mathbf{A}_k\mathbf{f}_i, \quad \mathbf{c}_i = \mathbf{A}_c\mathbf{f}_i, \quad i = 0, ..., M-1. \tag{3}$$

Here, $\mathbf{k}_i$ serves as the query representation, while $\mathbf{c}_i$ captures the content representation. And we compute the relationship matrix B as follows:

$$B = Q^T V, \quad Q = \left[\mathbf{k}_0\ \mathbf{k}_1\ ...\ \mathbf{k}_{M-1}\right], \quad V = \left[\mathbf{c}_0\ \mathbf{c}_1\ ...\ \mathbf{c}_{M-1}\right]. \tag{4}$$

The resulting sequence representation $\mathbf{B}$ maintains a consistent dimension of $[D \times D]$ **regardless of sequence length**. Finally, the sequence classification score is computed as:

$$Output_{seq}(T) = \mathbf{V}_2\text{Conv1D}(B). \tag{5}$$

where $\mathbf{V}_2$ is a weight vector applied after convolution.

We utilize the outputs from both the single-frame classifier $Output_{single}(T)$ and the sequence classifier $Output_{seq}(T)$ to simultaneously supervise the model's learning. We used the diagnostic information from a single image as the label for the single-frame classifier and the final prediction result as the label for the sequence classifier. The final loss function is defined as:

$$\text{loss}_{\text{final}} = \lambda_1 \text{loss}_{\text{single}} + \lambda_2 \text{loss}_{\text{seq}}, \quad (6)$$

where $\text{loss}_{\text{single}}$ and loss_{seq} are the cross-entropy losses from the single-frame classifier and sequence classifier, respectively. We use the coefficients λ_1 and λ_2 to control the contribution of these two components to the total loss function.

3 Experiment

3.1 Datasets and Implementation Details

The OHTS [5,10] dataset represents a landmark multicenter clinical investigation conducted across 22 sites in 16 U.S. states to investigate the progression to primary open-angle glaucoma (POAG). The study enrolled 1,636 participants aged 40–80 years, with intraocular pressure readings of 24-32 mmHg in one eye and 21-32 mmHg in the contralateral eye. Annual color fundus photography was performed, with POAG status assessments conducted at a centralized Optic Disc Reading Center using a standardized evaluation protocol. The assessment methodology involved independent evaluations by two masked certified readers, with discordant cases adjudicated by a senior masked reader. A quality assurance analysis of 86 eyes demonstrated substantial inter-rater reliability ($\kappa = 0.70$, 95% CI: 0.55–0.85). Although annual follow-up was targeted, actual visit intervals varied based on participant availability. We utilized 37,399 fundus images, along with corresponding temporal data (in yearly intervals) and POAG diagnostic outcomes for analysis. We excluded terminal visit images for each eye, retaining only the initial image in cases of multiple same-visit acquisitions. The resulting cohort included 30,932 images from 1,597 participants. The dataset maintains a realistic distribution of negative cases, comprising 20% of the cohort across training, testing, and validation sets, reflecting real-world clinical proportions.

The GRAPE [12] dataset encompasses 1,115 clinical encounters derived from 263 eyes of 144 patients with confirmed glaucoma diagnoses. Longitudinal follow-up included 3–9 visits per eye, with a minimum inter-visit interval of 5 months. The study population has a mean age of 42.49 years, with a balanced distribution across gender and ocular laterality. We perform visual field (VF) progression prediction on the GRAPE dataset. Several approaches exist for identifying glaucoma progression in visual fields [19,23]. This dataset incorporates three widely adopted automated assessment methods optimized for Octopus perimetry data. Two methods utilize point-wise linear regression (PLR) analysis, where progression is confirmed if either two (PLR2) or three (PLR3) points exhibit significant negative slopes ($P < 0.01$). The third method assesses mean deviation (MD) trends over time, identifying progression when a statistically significant negative MD slope is detected ($P < 0.05$).

Data Preprocessing and Fairness Considerations. We applied unified preprocessing to the dataset to ensure fair comparison. Specifically, for samples that were later converted to positive, we only used their earlier negative samples to ensure that no glaucoma-positive samples were present in the input images. For both baseline methods and our approach, we used identical preprocessing protocols, inputting the temporally last single image, which ensures consistency in the interval between model input and prediction time points, thereby guaranteeing the fairness of results. We selected LSTM and ResNet as baseline methods because current state-of-the-art approaches (e.g., DeepGF [15], GLIM-Net [11]) require additional information inputs (such as polar-transformed fundus images or attention maps) that our model does not need. These input disparities would prevent fair comparison.

TSDF has two core components: the feature representation module and the temporal aggregator module. **Feature Representation Module.** We employed the AdamW optimizer ($\beta_1 = 0.9$, $\beta_2 = 0.95$) with weight decay of 0.05, except for bias and normalization layers. The base learning rate was set to 1.5×10^{-4}, scaled linearly based on batch size (normalized to 256), with a 40-epoch warmup. Training lasted 400 epochs total. The input images were resized to 224×224 pixels with data augmentation including random cropping (scale range 0.2–1.0, bicubic interpolation) and horizontal flipping. Images were normalized using ImageNet statistics (mean = [0.485, 0.456, 0.406], std = [0.229, 0.224, 0.225]). We used ViT-Base [3] as the backbone with a 16×16 patch size and 75% masking ratio.

Temporal Aggregator Module. The temporal aggregator was implemented in PyTorch with AdamW optimizer ($\beta_1 = 0.5, \beta_2 = 0.9$) and trained using cross-entropy loss. The initial learning rate was 10^{-4} with cosine annealing scheduling decaying to 5×10^{-5}. Model weights were orthogonally initialized for all linear and convolutional layers. The feature dimension was set to 768. Training lasted a maximum of 50 epochs with early stopping if no improvement occurred for 10 consecutive epochs. Model selection was based on validation accuracy, with a weight decay of 10^{-3} maintained throughout the training process.

We compared our method against two baselines: a ResNet50-based model [7] and an LSTM-based temporal model [9]. The ResNet50 model, pre-trained on ImageNet, was fine-tuned for glaucoma prediction using single fundus images. In contrast, the LSTM model required a fixed-length sequence of Δt fundus images, using ResNet50 as a feature extractor, followed by bidirectional LSTM layers for temporal modeling. This design caused sample loss when patients had fewer than Δt images. To assess the impact of sequence length, we tested LSTM performance with Δt ranging from 1 to 4 for OHTS and 1 to 2 for GRAPE, as GRAPE samples generally contain shorter temporal sequences.

In this study, we used 5-fold random cross-validation in all experiments. In each cross-validation, patients were stratified into training (70%), validation (10%), and test (20%) sets at the participant level. Glaucoma prognosis performance was evaluated using accuracy (ACC) and area under the ROC curve

(AUC). Statistical significance was assessed using the Mann–Whitney U test, with a p-value indicating the probability of observing the difference between models under the null hypothesis. All experiments were conducted on NVIDIA GPUs with CUDA acceleration, implemented in PyTorch with gradient clipping at 5.0 for stable training.

3.2 Main Results

Quantitative Comparison on OHTS Dataset. As shown in Table 1, ResNet50 achieved the lowest performance metrics (ACC=0.88, AUC=0.78), primarily due to insufficient information from a single image for effective glaucoma prognosis. For the LSTM temporal model, increasing Δt improved both AUC and ACC (ACC increased from 0.86 to 0.89, AUC from 0.70 to 0.83), while significantly reducing usable sample size (from 3167 to 2763). Our approach significantly outperforms the baselines, achieving the highest AUC of 0.928 and ACC of 0.90 ($p < 0.01$). This improvement stems from its variable-length design, which utilizes all available samples for decision-making. Additionally, our method employs a decoupling strategy, resulting in a highly efficient model with only 4.1 million parameters—one-sixth the size of ResNet50 (23.5 million parameters) and LSTM (25.9 million parameters). This lightweight design enhances training efficiency and simplifies deployment.

Table 1. The results on the OHTS dataset

	Δt	ACC	AUC	PARAMs	Patient Num.
Resnet50	–	0.88	0.78	23,512,130	3167
LSTM	1	0.861	0.704	25,869,890	3167
	2	0.884	0.834		3038
	3	0.892	0.759		2888
	4	0.898	0.863		2763
TSDF	–	**0.907**	**0.931**	**4,136,452**	**3167**

Table 2. The results on the GRAPE dataset.

	Δt	ACC			AUC			Patient Num.
		PLR2	PLR3	MD	PLR2	PLR3	MD	
Resnet50	–	0.750	0.910	0.810	0.710	0.800	0.730	263
LSTM	1	0.774	0.925	0.943	0.733	0.744	0.519	263
	2	0.775	0.945	0.905	0.481	0.666	0.882	196
TSDF	–	**0.896**	**0.951**	0.911	**0.866**	**0.956**	**0.917**	**263**

Quantitative Comparison on GPAPE Dataset. As shown in Table 2, our method significantly outperforms the baselines, achieving the highest AUCs of 0.896, 0.951, and 0.917 for the PLR2, PLR3, and MD tasks, and the highest ACCs of 0.896 and 0.951 for the PLR2 and PLR3 tasks ($p < 0.01$). These results demonstrate significant performance improvements compared to the baseline while maintaining a highly efficient parameter configuration.

OHTS is a large multicenter clinical trial (1,597 participants, 22 locations), while GRAPE is a real-world dataset with different diagnostic criteria and a smaller cohort (144 patients). GRAPE includes three progression labels derived from different visual field assessment methods (PLR2, PLR3, MD). Our model's consistent performance across both datasets demonstrates strong generalizability and robustness.

3.3 Ablation Study

The Ablation Study of Decoupling Design. To demonstrate the effectiveness of our decoupling design, which separates feature extraction from prognosis, we compared our method with an end-to-end model using ResNet50 as the feature encoder. All settings remained identical except for the feature encoder and training logic (end-to-end vs. decoupling). The end-to-end model achieved 0.891 in accuracy and 0.505 in AUC (OHTS), as well as 0.947 in accuracy and 0.771 in AUC (GRAPE-PLR2), which are notably lower than our method's 0.900 in accuracy and 0.930 in AUC (OHTS) and 0.950 in accuracy and 0.956 in AUC (GRAPE-PLR2). These results confirm that our approach, leveraging self-supervised training and decoupled distribution training, better captures effective information and significantly improves accuracy.

The Ablation Study of Temporal Aggregator Module Design. The temporal aggregator module was trained using both intermediate module outputs and final predictions as loss functions. Balancing their contributions to the overall loss is crucial. Therefore, we investigated the impact of the loss function coefficients λ_1 and λ_2, as detailed in Table 3. The configuration $\lambda_1 = 1.5$, $\lambda_2 = 1$ achieved the best AUC and ACC performance. On the OHTS dataset, it achieved an ACC of 0.891 and an AUC of 0.928, while on the GRAPE dataset, it achieved an ACC of 0.947 and an AUC of 0.956 for the PLR3 task.

Next, we investigated the attention layer designed to balance performance and parameter efficiency. A larger parameter size enhances model robustness but is harder to train on small datasets and increases computational costs. Conversely, a smaller parameter size improves training efficiency but may reduce accuracy on large datasets. As shown in Table 4, we explored three approaches: (1) setting the intermediate layer size to match the input size while maintaining the original attention design,(2) adjusting the intermediate layer to double the input size, and (3) adding a self-attention layer to better capture temporal dependencies. Ultimately, for glaucoma prediction, the design with twice the input size achieved the best performance while maintaining a balanced parameter size.

Table 3. The ablation study of λ_1 and λ_2 in loss function on OHTS and GRAPE datasets

λ_1: λ_2	OHTS		GRAPE (PLR3)		PARAMs
	ACC	AUC	ACC	AUC	
0:1	0.877	0.920	0.939	0.948	4,136,452
0.5:1	0.890	0.913	0.947	0.953	
1:0	0.579	0.585	0.425	0.426	
1:1	0.889	0.919	0.947	0.956	
1.5:1	**0.891**	**0.928**	**0.947**	**0.956**	

Table 4. The ablation study of the temporal aggregator module on OHTS and GRAPE datasets.

	OHTS		GRAPE (PLR3)		PARAMs
	ACC	AUC	ACC	AUC	
input size	0.888	0.927	**0.947**	0.955	**2,956,036**
2 * input size	**0.891**	**0.935**	**0.947**	**0.956**	4,136,452
self-attention	0.887	0.903	**0.947**	0.954	8,855,044

4 Conclusion

In this paper, we propose TSDF, a two-stage framework for variable-length glaucoma prognosis. Specifically, we decouple this task into two stages: The first stage employs a self-supervised masked autoencoder for feature representation, enabling joint training across multiple glaucoma datasets and improving feature learning for small datasets. The second stage introduces a temporal aggregation module with masked non-local attention layers, efficiently aggregating variable-length sequences. Its dual-path design captures both single-frame and time-series frame information. We validate our approach on two glaucoma datasets of different scales, OHTS and GRAPE, demonstrating superior performance.

Acknowledgement. This work was supported in part by the National Eye Institute under grant R21EY035296, and in part by the National Science Foundation under grant 2306556. The content is solely the responsibility of the authors and does not represent the official views of the National Institutes of Health.

References

1. Bourne, R.R.A., et al.: Causes of vision loss worldwide, 1990–2010: a systematic analysis. Lancet Glob. Health **1**(6), e339–e349 (2013)
2. De Vente, C., et al.: Airogs: artificial intelligence for robust glaucoma screening challenge. IEEE Trans. Med. Imaging **43**(1), 542–557 (2023)

3. Dosovitskiy, A., et al.: An image is worth 16x16 words: transformers for image recognition at scale. In: International Conference on Learning Representations (ICLR) (2021)
4. Fan, R., Bowd, C., Brye, N., Christopher, M., Weinreb, R.N., Kriegman, D.J., Zangwill, L.M.: One-vote veto: semi-supervised learning for low-shot glaucoma diagnosis. IEEE Trans. Med. Imaging **42**(12), 3764–3778 (2023)
5. Gordon, M.O., et al.: The ocular hypertension treatment study: baseline factors that predict the onset of primary open-angle glaucoma. Arch. Ophthalmol. **120**(6), 714–720 (2002)
6. He, K., Chen, X., Xie, S., Li, Y., Dollár, P., Girshick, R.: Masked autoencoders are scalable vision learners. In: Proceedings of the IEEE/CVF Conference on Computer Vision and Pattern Recognition. pp. 16000–16009 (2022)
7. He, K., Zhang, X., Ren, S., Sun, J.: Deep residual learning for image recognition. In: Proceedings of the IEEE Conference on Computer Vision and Pattern Recognition, pp. 770–778 (2016)
8. Hemelings, R., et al.: Accurate prediction of glaucoma from colour fundus images with a convolutional neural network that relies on active and transfer learning. Acta Ophthalmol. **98**(1), e94–e100 (2020)
9. Hochreiter, S., Schmidhuber, J.: Long short-term memory. Neural Comput. **9**(8), 1735–1780 (1997)
10. Holste, G., et al.: Harnessing the power of longitudinal medical imaging for eye disease prognosis using transformer-based sequence modeling. NPJ Digital Med. **7**(1), 216 (2024)
11. Hu, X., et al.: Glim-net: chronic glaucoma forecast transformer for irregularly sampled sequential fundus images. IEEE Trans. Med. Imaging **42**(6), 1875–1884 (2023)
12. Huang, X., et al.: Grape: A multi-modal dataset of longitudinal follow-up visual field and fundus images for glaucoma management. Scientific Data **10**(1), 520 (2023)
13. Kamal, M.S., Dey, N., Chowdhury, L., Hasan, S.I., Santosh, K.: Explainable AI for glaucoma prediction analysis to understand risk factors in treatment planning. IEEE Trans. Instrum. Meas. **71**, 1–9 (2022)
14. Khalil, T., Khalid, S., Syed, A.M.: Review of machine learning techniques for glaucoma detection and prediction. In: 2014 Science and Information Conference, pp. 438–442. IEEE (2014)
15. Li, L., Wang, X., Xu, M., Liu, H., Chen, X.: DeepGF: Glaucoma forecast using the sequential fundus images. In: Martel, A.L., Abolmaesumi, P., Stoyanov, D., Mateus, D., Zuluaga, M.A., Zhou, S.K., Racoceanu, D., Joskowicz, L. (eds.) MICCAI 2020. LNCS, vol. 12265, pp. 626–635. Springer, Cham (2020). https://doi.org/10.1007/978-3-030-59722-1_60
16. Lin, M., et al.: Automated diagnosing primary open-angle glaucoma from fundus image by simulating human's grading with deep learning. Sci. Rep. **12**(1), 14080 (2022)
17. Lin, M., et al.: Multi-scale multi-structure siamese network (mmsnet) for primary open-angle glaucoma prediction. In: International Workshop on Machine Learning in Medical Imaging, pp. 436–445. Springer (2022)
18. Luo, Y., Shi, M., Tian, Y., Elze, T., Wang, M.: Harvard glaucoma detection and progression: a multimodal multitask dataset and generalization-reinforced semi-supervised learning. In: Proceedings of the IEEE/CVF International Conference on Computer Vision, pp. 20471–20482 (2023)

19. Saeedi, O.J., et al.: Agreement and predictors of discordance of 6 visual field progression algorithms. Ophthalmology **126**, 822–828 (2019). https://doi.org/10.1016/j.ophtha.2019.01.029
20. Singh, L.K., Khanna, M., et al.: A novel multimodality based dual fusion integrated approach for efficient and early prediction of glaucoma. Biomed. Signal Process. Control **73**, 103468 (2022)
21. Singh, L.K., Pooja, Garg, H., Khanna, M.: Deep learning system applicability for rapid glaucoma prediction from fundus images across various data sets. Evolving Syst. **13**(6), 807–836 (2022)
22. Tham, Y.C., Li, X., Wong, T.Y., Quigley, H.A., Aung, T., Cheng, C.Y.: Global prevalence of glaucoma and projections of glaucoma burden through 2040: a systematic review and meta-analysis. Ophthalmology **121**(11), 2081–2090 (2014)
23. Vesti, E., Johnson, C.A., Chauhan, B.C.: Comparison of different methods for detecting glaucomatous visual field progression. Invest. Ophthalmol. Vis. Sci. **44**, 3873–3879 (2003). https://doi.org/10.1167/iovs.02-1171
24. Zhang, L., Tang, L., Xia, M., Cao, G.: The application of artificial intelligence in glaucoma diagnosis and prediction. Frontiers Cell Dev. Biol. **11**, 1173094 (2023)
25. Zhou, Y., Yang, G., Zhou, Y., Ding, D., Zhao, J.: Representation, alignment, fusion: a generic transformer-based framework for multi-modal glaucoma recognition. In: International Conference on Medical Image Computing and Computer-Assisted Intervention, pp. 704–713. Springer (2023)

AI-Based Response Assessment and Prediction in Longitudinal Imaging for Brain Metastases Treated with Stereotactic Radiosurgery

Lorenz Achim Kuhn[1(✉)], Daniel Abler[2,3,4], Jonas Richiardi[4], Andreas F. Hottinger[3,4,7], Luis Schiappacasse[4,6], Vincent Dunet[5], Adrien Depeursinge[1,4,6], and Vincent Andrearczyk[1,4,6]

[1] Institute of Informatics, HES-SO Valais-Wallis, Sierre, Switzerland
{lorenz.kuhn,adrien.depeursinge}@hevs.ch

[2] Department of Oncology, Geneva University Hospitals (HUG), Geneva, Switzerland

[3] Department of Oncology, Lausanne University Hospital and University of Lausanne (UNIL), Lausanne, Switzerland

[4] Lundin Family Brain Tumour Research Centre, CHUV and UNIL, Lausanne, Switzerland

[5] Department of Medical Radiology, Service of Diagnostic and Interventional Radiology, Neuroradiology Unit, CHUV and UNIL, Lausanne, Switzerland

[6] Department of Oncology, Service of Nuclear Medicine, CHUV and UNIL, Lausanne, Switzerland

[7] Department of Medical Radiology, Service of Nuclear Medicine and Department of Neuroscience, CHUV, Lausanne, Switzerland

Abstract. Brain Metastases (BM) are a large contributor to mortality of patients with cancer. They are treated with Stereotactic Radiosurgery (SRS) and monitored with Magnetic Resonance Imaging (MRI) at regular follow-up intervals according to treatment guidelines. Analyzing and quantifying this longitudinal imaging represents an intractable workload for clinicians. As a result, follow-up images are not annotated and merely assessed by observation. Response to treatment in longitudinal imaging is being studied, to better understand growth trajectories and ultimately predict treatment success or toxicity as early as possible. In this study, we implement an automated pipeline to curate a large longitudinal dataset of SRS treatment data, resulting in a cohort of 896 BMs in 177 patients who were monitored for >360 days at approximately two-month intervals at Lausanne University Hospital (CHUV). We use a data-driven clustering to identify characteristic trajectories. In addition, we predict 12 months lesion-level response using classical as well as graph machine learning Graph Machine Learning (GML). Clustering revealed 5 dominant growth trajectories with distinct final response categories. Response prediction reaches up to 0.90 AUC (CI95% = 0.88–0.92) using only pre-treatment and first follow-up MRI with gradient boosting. Similarly, robust predictive performance of up to 0.88 AUC (CI95% = 0.86–0.90) was obtained using GML, offering more flexibility with a single model for multiple input time-points configurations. Our results

B. Hou and T. S. Mathai (Eds.): LMID 2025, LNCS 16184, pp. 102–112, 2026.
https://doi.org/10.1007/978-3-032-16128-4_10

suggest potential automation and increased precision for the comprehensive assessment and prediction of BM response to SRS in longitudinal MRI. The proposed pipeline facilitates scalable data curation for the investigation of BM growth patterns, and lays the foundation for clinical decision support systems aiming at optimizing personalized care.

Keywords: Longitudinal MRI · Brain Metastases · Response Prediction · Tumor Growth Trajectory · Graph Machine Learning

1 Introduction

BM represent a major cause of mortality in patients with cancer, occurring via the seeding of circulating tumor cells into the brain's microvasculature [1,13]. The most common primaries include lung (20–56%), melanoma (5–20%) and breast (7–16%) [1]. While systemic therapies and immune checkpoint inhibition offer therapeutic options, the blood-brain barrier limits their effectiveness [21].

Among current treatment modalities, radiation therapy, especially SRS, is preferably used when the lesion load is appropriate according to treatment guidelines, due to its noninvasive, targeted approach. Unlike whole-brain radiotherapy (WBRT), which affects healthy tissue and is used primarily in palliative care, SRS precisely targets individual lesions, allowing for repeated treatments with minimal damage to surrounding brain tissue [12]. This makes it particularly suitable for personalized healthcare applications.

Evaluating treatment response is crucial for adapting patient management strategies. The Response Assessment in Neuro-Oncology for Brain Metastases (RANO-BM) criteria [13] provide a widely used framework, classifying tumor evolution based on lesion diameter or, more recently, volumetric measurements [3,7,17,18]. Volumetric analysis is emerging as a more specific indicator for assessing progression or regression. Similar to [3], we adapt the volumetric RANO-BM assessment criteria to the lesion-level with response classes defined as: **Complete Response (CR):** complete disappearance of the tumor, **Partial Response (PR):** volume $<$ 34.3% of baseline, **Stable Disease (SD):** volume within 34.3%–172.8% of prior minimum, **Progressive Disease (PD):** volume $>$ 172.8% of prior minimum.

Understanding the dynamics of BM evolution through longitudinal imaging is key for developing predictive models. However, the natural development of untreated metastases is poorly characterized due to ethical constraints. Kobets et al. [12] have leveraged rare pre-treatment imaging to analyze growth rates, reporting that faster-growing tumors are associated with poorer outcomes. Other work has documented nonlinear growth patterns and phenomena such as pseudo-progression, an apparent lesion enlargement due to treatment-induced radiation necrosis rather than true progression [16,19].

Efforts to predict treatment outcomes have increasingly employed machine learning Machine Learning (ML) techniques. Kanakarajan et al. [11] combined clinical, radiomic, and deep features from manually annotated images to predict patient-level lesion control post-SRS for 129 patients with an accuracy of 0.82. They found that the addition of deep features did not significantly aid in prediction performance. Cho et al. [6] demonstrated that incorporating follow-up MRI studies improves predictive performance (Area Under the ROC Curve (AUC) $\approx$ 0.88). Cao et al. [5] introduced tumor-connectomics, a graph-based model, to distinguish true progression from radiation necrosis with high sensitivity and specificity.

Despite promising results, most of these studies rely on limited datasets, particularly in terms of number of lesions. The BraTS-METS 2023 Challenge highlighted the difficulty of accurate BM segmentation, with the best method achieving only 0.65 ± 0.25 Dice coefficient [14]. Low segmentation performance is associated with challenges, such as false negatives for small lesions. Resegmentation [2] is a strategy particularly well-suited for longitudinal image analysis, where a previously obtained (and potentially outdated) segmentation mask is provided as an auxiliary input to guide the propagation of segmentations to the current image context. Resegmentation-based methods [2] and longitudinal tracking via AI systems like METRO [10], are promising directions for robust and automated follow-up modeling.

SRS-based clinical workflows naturally generate large amounts of longitudinal imaging and segmentation data, which remain underutilized. We analyze a cohort of 896 BM from 177 patients treated with SRS at CHUV hospital. We implement an automated pipeline for processing MRI, Computed Tomography (CT) images and Radiotherapy Structure (RTStruct) data to extract lesion-level time series. We then analyze volumetric growth trajectories using clustering, and leverage time point features in Gradient Boosting and GML for response prediction.

This data-centric approach aims to address current limitations in sample size and longitudinal modeling, providing a robust foundation for treatment response prediction and biomarker discovery in BM management.[1]

2 Materials and Methods

2.1 Data

A large dataset consisting of 1135 individual patients with BM, treated using SRS, was extracted and depersonalized from the Picture Archiving and Communication System (PACS) at CHUV hospital as approved by the Research Ethics Committee of Vaud Canton, Switzerland (No. 2024-00100). The data amounted to approximately 6.3 terabytes of DICOM files, with multiple imaging modalities including MRI sequences, CT scans, and RTStruct containing BM Region of Interest (ROI). These spanned multiple imaging sessions over each patient's clinical observation period. In total, 42,321 image sequences from 14,346 image acquisition sessions were extracted, averaging approximately 13 sessions per patient. The curation pipeline is depicted in Fig. 1 and detailed below.

[1] The code for the curation pipeline and the analysis part are available on https://github.com/Bangulli/BMPipeline and https://github.com/Bangulli/BMDataAnalysis.

Raw DICOM files were filtered by automatically selecting useful images and ROI, and converting sequences of interest, as defined by the DICOM metadata (by SequenceName, ContrastBolusAgent, and SeriesDescription), into a homogeneous data structure similar to BIDS [8]. ROI were automatically matched to the appropriate MRI in the patient timeline, because, in clinical practice, delineations are made with MRI images as visual reference on treatment CT images and the link ROI-MRI is then lost. Longitudinal image series were cleaned up by filtering large time gaps ($\geq$ 4 months), redundancies and pre-treatment data. Spatial alignment was performed by registering the CT, on which the ROI is spatially defined, to the corresponding MRI as well as all subsequent MRIs and ROI to the treatment MRI (t_0). All registrations performed affine alignments with the ANTs package [4].

UNet-based re-segmentation [2] was used to propagate the delineations from SRS at t_0 to the patients' treatment and follow-up MRI to obtain homogeneous ROI for lesions at all time points. Note that this model achieved a Dice score of 0.78 on a separate test set as reported in [2]. Individual ROI were extracted from the time points and matched by maximum overlap and minimum centroid distance, to achieve lesion correspondence over time.

This process yielded a lesion-level dataset of 896 lesions in 177 patients due to the inclusion criteria for this study: (1) at least one RTStruct (containing BM ROI), (2) a maximum average follow-up interval of 90 d and (3) a minimum observation period of 300 d and a semi-automatic sample rejection based on suspicious trajectory detection (time point swings into CR) and marginal manual observation. This ensured that only patients with complete data availability for the analysis were selected. Its scale and fully automated curation distinguish this dataset from prior brain metastasis imaging resources.

2.2 Growth Trajectory Analysis

To identify characteristic lesion growth patterns, volumetric trajectories were clustered using normalized feature vectors, where each lesion's volume was divided by its initial value at t_0 to emphasize relative change over time. t_0 always corresponds to the lesion's initial SRS. Due to the irregular timing of follow-up scans (mean interval 60 ± 20 days), all trajectories were resampled to seven uniformly spaced time points (t_0 to t_6, 60-day intervals). Three resampling methods were evaluated: Nearest Neighbor (NN), linear, and B-spline. NN interpolation was ultimately selected for its simplicity and its preservation of correspondence to real scan dates. B-spline and linear interpolation were deemed less suitable in this context due to overfitting of splines and loss of image correspondence. Resampled feature vectors were clustered using StepMix [15] based on Gaussian mixture modeling. The target cluster count was fixed to 5, to both maintain interpretability and avoid fractured micro-clusters with low membership counts.

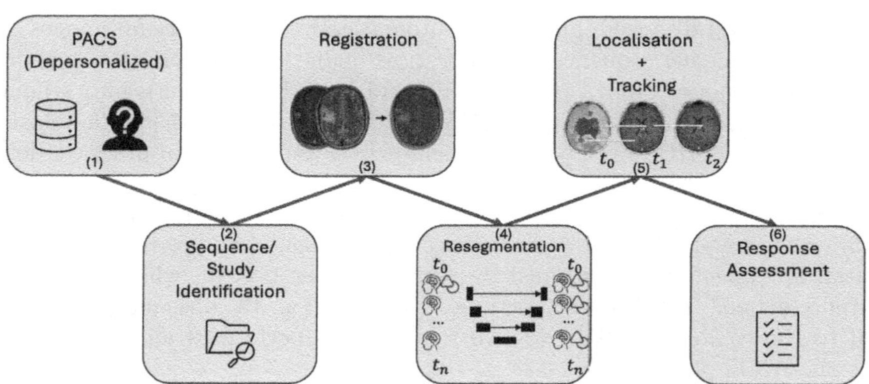

Fig. 1. Data curation pipeline. (1) Data is extracted and depersonalized from the PACS. (2) relevant DICOM files selection based on metadata filtering. (3) longitudinal imaging series are registered (first all ROI warped to their corresponding MRI using CT-MRI registration, then all MRI and ROIs to the t_0 MRI). (4) The tumor delineation for SRS at t_0 is propagated to the rest of the time series with UNet based re-segmentation. (5) Individual lesions are separated, localized and tracked. (6) The lesion response category is computed for each time point.

2.3 One-Year Response Prediction

Lesion-level response prediction after one-year was performed using clinical and radiomic features (including volume) extracted across resampled time points (as described in Sect. 2.2). Clinical variables included demographics, primary tumor site and histology, initial lesion volume, total lesion burden, BM location in brain, and lesion count. 107 radiomic features were extracted from the lesion ROI using PyRadiomics [9]. Radiomic and volume features at noisy time points (entering and immediately leaving CR, due to errors like incorrect MRI modality causing misdetection) were imputed using time-weighted linear interpolation. A value of zero was used for missing clinical variables. Categorical variables were one-hot encoded.

The prediction target was the lesion-level response category at t_6 (one-year). Two clinically meaningful binary classification tasks were defined: (1) CR vs. non-CR, and (2) responding {CR, PR} vs. non-responding {SD, PD}. Class distributions were 58.59% CR / 41.41% non-CR and 72.77% responding / 27.23% non-responding. Evaluation was performed with a 5-fold cross-validation and fold-wise feature standardization. Performance was assessed based on test observations pooled across the 5 folds, where a permutation test with 1000 iterations was used to assess the significance of the performance difference.

Classical ML. Feature vectors were formed by concatenating per-time-point features to capture longitudinal data. Predictions were done using Light Gradient Boosting Machine (LGBM), trained with class weight balancing. Models were

Table 1. Lesion-level characteristics.

Category	Metric	Female (n = 479)	Male (n = 417)
Initial volume (mm^3)	Mean	565.02	402.17
	Median	53.00	64.00
	Std Dev	2032.77	1161.08
Lesion one-year response (RANO-BM)	CR (Complete Response)	295	230
	PR (Partial Response)	56	71
	SD (Stable Disease)	65	31
	PD (Progressive Disease)	63	85
Primary cancer type (n)	Lung	219	218
	Melanoma	91	98
	Breast	100	–
	Unknown	36	36
	Kidney	20	25
	Epithelial tumor	–	27
	Esophagus	8	3
	Other	5	10

trained on combinations of time points (t_0, $t_0\&t_1$, $t_0\&t_1\&t_2$, etc.) to assess at which follow up time the one-year response can be predicted reliably.

GML. Graph models were used to capture temporal dependencies in lesion evolution. Each lesion trajectory was modeled as a graph where nodes represent time points (t_0 to t_5) and edges encode temporal relationships via normalized time deltas.

Multiple graph configurations were tested, including dense/sparse and directed/undirected graphs. A fully connected and directed (to past) graph was selected as optimal based on volume-only baselines. The Graph Attention Network (GAT) [20] was selected as encoder architecture for its capability to learn attention between time points in the given graph configuration. Graphs were processed using PyTorch Geometric (PyG)[2].

Prediction was framed as a whole-graph classification task. Two training modes were evaluated: (1) time-specific models trained on fixed graph configurations, i.e. the models were trained to specifically predict for one fixed time point configuration (similar to the classical ML models), and (2) general models trained with randomly cropped graphs to achieve generalizability over time point configurations. Models consisted of a shallow single-layer GAT encoder and a linear classification head. The Adam optimizer with a cosine annealing scheduler was used with an initial learning rate of 0.0001 with warm restarts every 50 epochs. Models were trained up to 1000 epochs with early stopping (patience = 20), using class-balanced binary cross-entropy loss.

[2] https://pytorch-geometric.readthedocs.io.

Table 2. ROC AUC with 95% Confidence Intervals for each method and configuration. $t_0{:}t_N \rightarrow t_6$ signifies prediction of response at t_6 using all time points from t_0 to t_N

Method	$t_0 \rightarrow t_6$	$t_0{:}t_1 \rightarrow t_6$	$t_0{:}t_2 \rightarrow t_6$	$t_0{:}t_3 \rightarrow t_6$	$t_0{:}t_4 \rightarrow t_6$	$t_0{:}t_5 \rightarrow t_6$
CR vs non-CR						
GML general	0.70 [0.66–0.73]	0.88 [0.86–0.90]	0.91 [0.89–0.93]	0.94 [0.92–0.95]	0.96 [0.95–0.97]	0.98 [0.98–0.99]
GML time-specific	0.76 [0.73–0.79]	0.88 [0.86–0.91]	0.92 [0.89–0.93]	0.94 [0.92–0.95]	0.97 [0.95–0.98]	0.98 [0.97–0.99]
Classic	0.77 [0.74–0.81]	0.90 [0.88–0.92]	0.93 [0.91–0.95]	0.95 [0.93–0.96]	0.97 [0.96–0.98]	0.99 [0.98–0.99]
Responding vs non-Responding						
GML general	0.58 [0.54–0.62]	0.78 [0.74–0.81]	0.81 [0.79–0.84]	0.84 [0.82–0.87]	0.87 [0.85–0.89]	0.90 [0.88–0.92]
GML time-specific	0.61 [0.57–0.66]	0.78 [0.76–0.81]	0.82 [0.79–0.84]	0.85 [0.82–0.87]	0.87 [0.85–0.89]	0.90 [0.88–0.92]
Classic	0.61 [0.57–0.65]	0.82 [0.79–0.85]	0.87 [0.85–0.89]	0.90 [0.88–0.92]	0.93 [0.92–0.95]	0.97 [0.96–0.98]

3 Results and Discussion

3.1 Data Curation

A dataset of 896 individual BM from 177 patients was obtained from the data curation pipeline. Table 1 provides a lesion-level summary of initial volumes, one-year response category distributions and primaries. The evolution sequence of the responses over the observation span is shown in Fig. 2. Dominant volume evolution trajectories revealed with clustering (Sect. 2.2) are depicted in Fig. 3.

While manual processing of such quantity of data is unfeasible, the proposed semi-automatic pipeline allows us to leverage the wealth of longitudinal data, paving the way to clinical assistance in making informed decisions as well as generating knowledge on lesion growth trajectories. A representative quality control study on the individual time-point level, conducted by L.A.K. (computer scientist), showed that mis-segmentations (approx. 16% partial segmentation and other failures) and mis-registrations (approx. 1% misaligned image pairs) may occur on rare occasions. The quality control was done by qualitative evaluation of a random subset of 66 lesion time series. Automated rejection of outliers (suspicious growth trajectories with longer swings into CR) reduced the need for manual cleaning to a minimal level. For many patients (76.5%), RTStruct files were stored on separate databases that cannot be queried. Missing data will be re-extracted for larger analyses in future work.

3.2 Lesion Growth Trajectories

The response flow diagram shown in Fig. 2 shows that most lesions will achieve CR after one year. It is worth noting that more than 50% of the lesions resulting

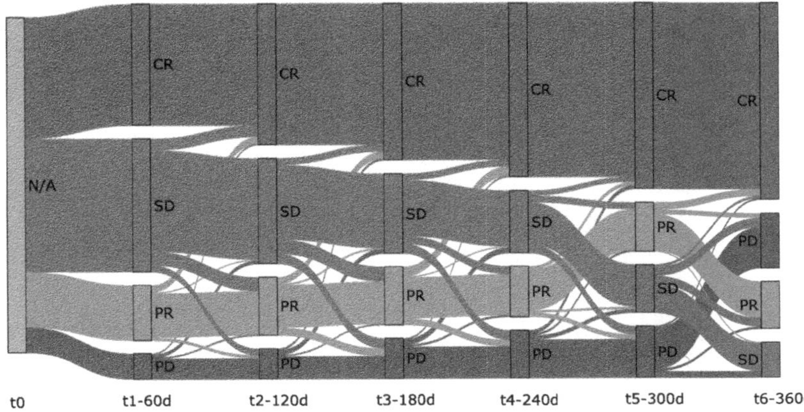

Fig. 2. Sankey flow diagram of lesion response categories over time. Imaging time points t_i are spaced 60 days apart. t_0 is the treatment day.

in CR already enter CR at the first follow-up (t_1, 60 days post treatment). Furthermore a large portion of lesions appear as SD in early follow-ups but separates into CR and the partial categories later. Clustering lesion volume trajectories over time (Sect. 2.2) revealed five characteristic growth patterns, as depicted in Fig. 3. Each of the five clusters exhibited a distinct volumetric profile. *Cluster 0* shows a consistent and strong volumetric reduction, indicative of CR. This trajectory corresponds to the lesions in the flow diagram (Fig. 2) that enter CR in the early time points. *Cluster 1* displays an initial increase in volume followed by reduction, a profile suggestive of pseudoprogression. *Cluster 2* includes lesions that respond to treatment with moderate shrinkage, though not to the extent of a complete response. This cluster has the most random outcome, as shown by the relatively even distribution of response categories at t_6, indicated in the corresponding pie chart in Fig. 3 and comprises largely of lesions that spend the early follow-up time in the SD category in the flow diagram (Fig. 2). *Cluster 3* comprises a small number of lesions with rapid volumetric increase, suggesting aggressive malignant progression. However, this small cluster may be affected by inconsistencies in terms of segmentation and registration errors introduced by the processing pipeline. *Cluster 4* is characterized by gradual but accelerating volume increase, consistent with PD. These clusters highlight distinct volumetric behavior over time and may support clinical interpretation of treatment efficacy.

3.3 Response Prediction

The first task was to predict CR/non-CR one year after treatment. The AUC are reported in Table 2 with the 95% Confidence Interval (CI) obtained by bootstrapping the prediction results of all folds. The second task was to predict whether a lesion will be responding or non-responding one year after treatment. Table 2 summarizes prediction performance based on AUC, also with 95% CI.

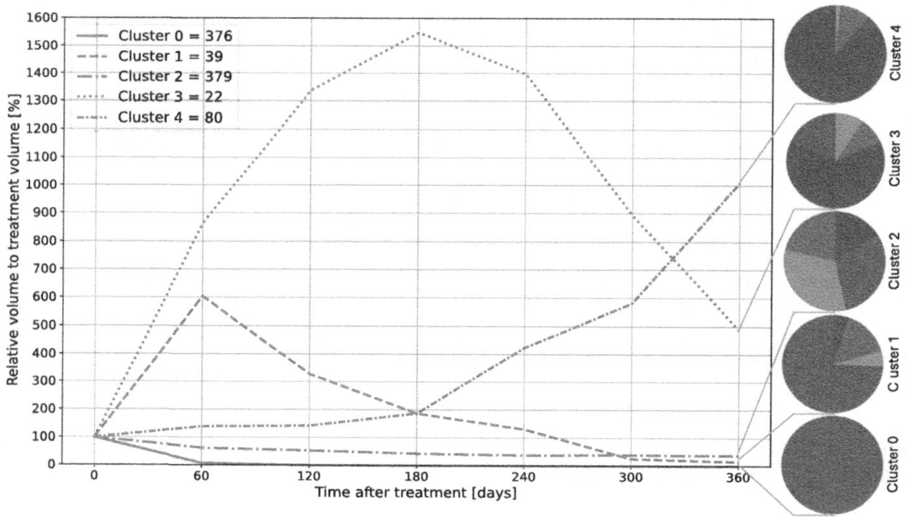

Fig. 3. Dominant lesion growth trajectories revealed by StepMix with a target of 5 clusters. Trajectories are computed as the mean of all cluster members at each time point. The one year response distribution of each cluster is given by the pie charts, where the response categories include CR, PR, SD, and PD. Response categories were assigned based on the criteria given in Sect. 1, using the volume of the resegmented ROI.

All prediction models achieved robust performance based on pre-treatment t_0 and first follow-up t_1 (see Table 2). Predicting CR seems to be a simpler task, with respect to the response prediction, indicated by better performance across the board.

In the CR prediction task, the LGBM method slightly outperforms the GML configurations (Table 2). However, the permutation test showed that this improvement only reached significance ($p < 0.05$) compared to the general GML model when data from only t_0 are used, and the performance gain over the time-specific model is consistently insignificant ($p > 0.05$).

In the response prediction task, the LGBM model achieves the highest performance scores (Table 2) when compared to the GML approaches. The performance gain becomes significant ($p < 0.05$) when longitudinal data are used, at the cost of reduced flexibility (one model per time-points configuration).

Performance improves significantly with each additional follow-up time-point included in the model ($p < 0.05$) for both prediction tasks, with the highest AUC achieved when using the full follow-up period (t_0:$t_5 \rightarrow t_6$). Notably, the most substantial gain occurs when the first post-SRS follow-up (t_1) is used.

The flow diagram in Fig. 2 reveals that most of the lesions achieving CR after one year are already in CR at the first follow-up. This pattern could contribute to the relatively higher separability of CR vs. non-CR in predictive models, as compared to Responding vs. non-Responding.

4 Conclusion

We presented an automatic pipeline for curating longitudinal BM imaging data resulting from routine SRS. The curated data allowed investigating lesion growth trajectories, where clustering revealed five dominant characteristic temporal patterns that could be used to guide early response identification. We also investigated one-year response prediction at lesion-level using classical and graph ML. Longitudinal information (i.e. using multiple time points from t_0 to t_5) significantly improved predictive performance compared to using only a single time point t_0. This prediction yields robust results as soon as the first follow-up time point t_1 at 60 days post treatment is introduced, which may assist clinicians in making informed decisions in personalized healthcare for further treatment planning.

Limitations and Future Work: Our work contains several limitations and future work directions. First, trajectory analysis was pooling BMs from various primary cancers. Primary-specific clustering will be investigated in future work. Classical models are time-invariant, and current GML models do not fulfill their expectation of performance gain. Future work includes reformatting the task as node-level prediction and extending graph models to include richer clinical or spatial information and encoding the target time into the prediction task. Preliminary regression experiments to predict the tumor volume itself at the one-year follow-up showed promise but were biased by the over-representation of resolved lesions. A two-step model—classification followed by regression—may better handle this imbalance.

Acknowledgments. This work was partially funded by the Swiss Cancer Research foundation with the project TARGET (KFS-5549-02-2022-R), the Lundin Family Brain Tumour Research Centre at CHUV, and the Swiss National Science Foundation (SNSF) with the project number 205320_219430.

Disclosure of Interests. No author has any conflict of interests to report.

References

1. Achrol, A.S., et al.: Brain metastases. Nat. Rev. Dis. Primers. **5**(1), 5 (2019). https://doi.org/10.1038/s41572-018-0055-y
2. Andrearczyk, V., et al.: Automatic detection and multi-component segmentation of brain metastases in longitudinal MRI. Sci. Rep. **14**(1), 31603 (2024)
3. Andrearczyk, V., et al.: The value of AI for assessing longitudinal brain metastases treatment response. Neuro-Oncol. Adv. **7**(1), vdae216 (2025). https://doi.org/10.1093/noajnl/vdae216
4. Avants, B.B., et al.: Advanced normalization tools (ants). Insight J. **2**(365), 1–35 (2009)
5. Cao, Y., et al.: A multidimensional connectomics- and radiomics-based advanced machine-learning framework to distinguish radiation necrosis from true progression in brain metastases. Cancers **15**(16) (2023)

6. Cho, S.J.e.a.: Prediction of treatment response after stereotactic radiosurgery of brain metastasis using deep learning and radiomics on longitudinal MRI data. Sci. Rep. **14**(1), 11085 (2024). https://doi.org/10.1038/s41598-024-60781-5
7. Douri, K., et al.: Response assessment in brain metastases managed by stereotactic radiosurgery: a reappraisal of the RANO-BM criteria. Curr. Oncol. **30**(11), 9382–9391 (2023). https://doi.org/10.3390/curroncol30110679
8. Gorgolewski, K.J., et al.: The brain imaging data structure, a format for organizing and describing outputs of neuroimaging experiments. Sci. Data **3**(1), 160044 (2016). https://doi.org/10.1038/sdata.2016.44
9. van Griethuysen, J.J., et al.: Computational radiomics system to decode the radiographic phenotype. Can. Res. **77**(21), e104–e107 (2017). https://doi.org/10.1158/0008-5472.CAN-17-0339
10. Hsu, D.G., et al.: Automatically tracking brain metastases after stereotactic radiosurgery. Phys. Imaging Radiat. Oncol. **27**, 100452 (2023). epub 2023 Jun 1. PMID: 37720463. PMCID: PMC10500025. https://doi.org/10.1016/j.phro.2023.100452
11. Kanakarajan, H., De Baene, W., Hanssens, P., Sitskoorn, M.: Predicting local control of brain metastases after stereotactic radiotherapy with clinical, radiomics and deep learning features. Radiat. Oncol. **19**(1), 182 (2024). https://doi.org/10.1186/s13014-024-02573-9
12. Kobets, A., Backus, R., Fluss, R., Lee, A., Lasala, P.: Evaluating the natural growth rate of metastatic cancer to the brain. Surg. Neurol. Int. 11, p. 254 (2020). https://doi.org/10.25259/SNI_291_2020
13. Lin, N.U., et al.: Response assessment criteria for brain metastases: proposal from the RANO group. Lancet Oncol. **16**(6), e270–e278 (2015). https://doi.org/10.1016/S1470-2045(15)70057-4
14. Moawad, A.W., et al.: The brain tumor segmentation - metastases (brats-mets) challenge 2023: Brain metastasis segmentation on pre-treatment mri. ArXiv [Preprint] (Dec 2024). arXiv:2306.00838v3, preprint published on arXiv. PMID: 37396600. PMCID: PMC10312806
15. Morin, S., et al.: Stepmix: a python package for pseudo-likelihood estimation of generalized mixture models with external variables (2023). arXiv preprint arXiv:2304.03853
16. Ocaña-Tienda, B., et al.: Growth dynamics of brain metastases differentiate radiation necrosis from recurrence. Neuro-Oncol. Adv. **5**(1), vdac179 (12 2022). https://doi.org/10.1093/noajnl/vdac179
17. Ocaña-Tienda, B., et al.: Volumetric analysis: rethinking brain metastases response assessment. Neuro-Oncol. Adv. **6**(1), vdad161 (12 2023). https://doi.org/10.1093/noajnl/vdad161
18. Oft, D., et al.: Volumetric regression in brain metastases after stereotactic radiotherapy: time course, predictors, and significance. Front. Oncol. **10** (2021)
19. Patel, T., McHugh, B., Bi, W., Minja, F., Knisely, J., Chiang, V.: A comprehensive review of MR imaging changes following radiosurgery to 500 brain metastases. Am. J. Neuroradiol. **32**(10), 1885–1892 (2011). https://doi.org/10.3174/ajnr.A2668
20. Veličković, P., Cucurull, G., Casanova, A., Romero, A., Liò, P., Bengio, Y.: Graph attention networks (2018)
21. Wilhelm, I., Molnár, J., Fazakas, C., Haskó, J., Krizbai, I.A.: Role of the blood-brain barrier in the formation of brain metastases. Int. J. Mol. Sci. **14**(1), 1383–1411 (2013). https://doi.org/10.3390/ijms14011383

Scalable Modeling of Nonlinear Network Dynamics in Neurodegenerative Disease

Daniel Semchin[1], Emile d'Angremont[2], Marco Lorenzi[3], and Boris A. Gutman[1(✉)]

[1] Illinois Institute of Technology, Department of Biomedical Engineering, Chicago, IL, USA
dsemchin@hawk.illinoistech.edu , bgutman1@iit.edu

[2] Department of Anatomy and Neurosciences, Amsterdam University Medical Center, Amsterdam, The Netherlands

[3] Epione Team, Inria Center of Université Côte d'Azur, Sophia Antipolis, Valbonne, France

Abstract. Mechanistic models of progressive neurodegeneration offer great potential utility for clinical use and novel treatment development. Toward this end, several connectome-informed models of neuroimaging biomarkers have been proposed. However, these models typically do not scale well beyond a small number of biomarkers due to heterogeneity in individual disease trajectories and a large number of parameters. To address this, we introduce the Connectome-based Monotonic Inference of Neurodegenerative Dynamics (COMIND). The model combines concepts from diffusion and logistic models with structural brain connectivity. This guarantees monotonic disease trajectories while maintaining a limited number of parameters to improve scalability. We evaluate our model on simulated data as well as on the Parkinson's Progressive Markers Initiative (PPMI) data. Our model generalizes to anatomical imaging representations from a standard brain atlas without the need to reduce biomarker number.

Keywords: Disease Progression Modeling · Brain Connectivity · Neurodegenerative Disease

1 Introduction

Recent advances in data-driven disease progression modeling (DPM) have significantly enhanced our understanding of neurodegenerative diseases. Such models promise broad applicability, both as simulators of treatment effect and to estimate pre-symptomatic time of disease onset. DPMs and related inference algorithms, i.e. methods inferring long-term biomarker trajectories from observations over a short time period, can be broadly classed along several dichotomies, e.g.: discrete- vs. continuous-time, mechanistic vs. purely generative statistical, monotonic vs. non-monotonic, subtyped vs. single-mode. Other ways to classify

B. Hou and T. S. Mathai (Eds.): LMID 2025, LNCS 16184, pp. 113–124, 2026.
https://doi.org/10.1007/978-3-032-16128-4_11

DPM models probably exist, but we find these four most useful for contextualizing this work. Discrete-time methods are largely derivates of the Event-Based Model (EBM) [1]. EBM assumes monotonic biomarker trajectories, described by an ordering of discrete stages. The method works well in cross-sectional datasets. A widely used subtyping derivative of EBM, Subtype and Stage Inference (SuStaIn) [2] seeks to disentangle temporal and phenotypic heterogeneity to identify population subgroups with common patterns of disease progression. The method groups individuals with common canonical biomarker orderings, defining disease stages independently for each subtype. A continuous-time alternative, the DIVE model [3] infers coherent spatiotemporal patterns using a parametric model of progression. DIVE both segments the cortex and finds cluster-level parameters of logistic progression curves, requiring longitudinal data for stable inference. Disease Progression Modelling and Stratification (DP-MoSt) [4] also fits logistic curves, simultaneously subtyping biomarker trajectories and samples into discrete patient groups. More recently, approaches like Brain Latent Progression (BrLP) [5] have integrated latent diffusion models with prior knowledge to enhance spatiotemporal disease progression predictions. DIVE and DP-MoSt both assume monotonic biomarker change. With the exception of DIVE, these approaches primarily operate on predefined regions of interest and no method above explicitly incorporates the brain's connectivity structure, limiting their ability to capture mechanistic aspects of disease spread. Complementing the subtyping approaches, network-based models have emerged as powerful tools for understanding disease progression through the lens of brain connectivity. These approaches are supported by neuropathological evidence for "prion-like" transsynaptic transmission of disease agents like misfolded tau and beta amyloid, suggesting that disease is transmitted along neuronal pathways rather than by proximity. An early example, the Network Diffusion Model (NDM) [6] models disease transmission as a diffusive mechanism over a tractography-based connectome. NDM demonstrated significantly higher accuracy in predicting longitudinal patterns of atrophy and metabolism compared to non-network-based models. Generalizing this, the Accumulation-Clearance-Propagation (ACP) [7] model expands the connectivity-based propagation idea with a three-compartment model allowing for time-varying biomarker diffusion effects that represent distinct processes. These and related [8] progression models explicitly use a system of differential equations; particularly when coupled with empirically established structural connectivity, the models provide a mechanistic explanation for disease spread in the brain. However, most current implementations are limited in their scalability to the full brain connectome with dozens or hundreds of regions.

As highlighted in recent reviews, data-driven disease progression models provide unique advantages over 'black box' machine learning tools by being inherently interpretable and requiring fewer data [9,10]. However, a critical gap remains in our current modeling arsenal: the lack of approaches that simultaneously leverage the brain's connectivity structure, scale to large numbers of brain regions, accommodate disease heterogeneity, and provide mechanistic interpretability over extended time ranges. Furthermore, the scale of modern neuroimaging datasets and high-resolution connectome mapping requires models that can efficiently operate across dozens to hundreds of brain regions while maintaining computational tractability. Existing approaches either sacrifice scale for mechanistic detail or interpretability for computational efficiency, limiting their practical application.

The approach described here - the Connectome-based Monotonic Inference of Neurodegenerative Dynamics (COMIND) - represents a step to address these requirements by developing a network-connectome based disease progression model that scales to large numbers of brain regions while providing interpretable, mechanistic insights into disease evolution over extended time periods. Our method improves scalability of network-based mechanistic modeling, offering new opportunities for understanding the complex spatiotemporal dynamics of neurodegenerative disease progression.

2 Methods

A number of the proposed progression models have used logistic evolution as the basic model-building element. The idea is appealing as it mirrors our intuition that neurodegeneration has a gradual onset, a period of rapid advance, and a saturation point. Indeed, the sigmoid is a solution to the logistic ODE $\dot{x} = k(1-x)x$, a dynamic in which the rate of accumulation is a product of self-supply and capacity. Our motivation is to incorporate network dynamics into this idea, while maintaining the asymptotic properties of the univariate logistic ODE.

2.1 Dynamic Propagation Model

We begin by considering a dynamic p-dimensional system of accumulating neurodegenerative effects with a static transition matrix $K \in \mathbb{R}^{p\times p}$ defined in some way by brain connectivity. As in a standard logistic differential equation, we define the rate of neurodegenerative accumulation as a product of the present state of neurodegeneration and remaining capacity. This can be expressed as

$$\frac{d\mathbf{x}}{dt} = [I - D(\mathbf{x})]K\mathbf{x}, \quad \mathbf{x} \in [0,1]^p \tag{1}$$

Here, $D(\mathbf{x})$ is the diagonal matrix with entries in $\mathbf{x}$. We consider external sources of neurodegeneration as constant, again contributing to the rate of regional accumulation in proportion to regional capacity.

$$\frac{d\mathbf{x}}{dt} = [I - D(\mathbf{x})][K\mathbf{x} + \mathbf{f}], \quad \mathbf{x} \in [0,1]^p, \quad \mathbf{f} \in [0,\infty)^p, \tag{2}$$

where $\mathbf{f}$ represents the constant "forcing" term in the ODE. For the kth individual regional biomarker, this is equivalent to $\dot{x}^k(t) = (1 - x^k)(\sum_m K_{km}x^m + f^k)$. To reduce the number of model parameters, here we consider the transition matrix simply as a scaled version of the connectome K^*, i.e. $K = s_t K^*$, where s_t is a parameter denoting timescale. Finally, to construct canonical trajectories of neurodegeneration we must scale solutions of the dynamic system to the natural scale of imaging biomarkers. This adds a scaling vector $\mathbf{s} \in \mathbb{R}^p$ to our trajectory parameter set. Suppose $\mathbf{x}(t)$ is the solution to (2). Then the predicted biomarker value at time t is $\hat{\mathbf{y}}(t) = \mathbf{s} \odot \mathbf{x}(t)$, where $\odot$ is element-wise multiplication. In summary, given a canonical connectome, our biomarker trajectory $\hat{\mathbf{y}}(t) = \hat{\mathbf{y}}(t;\theta)$ is parameterized by $\theta = \{s_t, \mathbf{s}, \mathbf{f}\}$, with $2p + 1$ parameters.

2.2 Modeling Subject-Specific Time-Shift

As is common practice, we assume a Gaussian i.i.d. noise model. For the kth biomarker y_{ij}^k measured in subject i at timepoint j, biomarker likelihood is expressed as

$$P(y_{ij}^k|\theta, \beta_i) = \mathcal{N}(y_{ij}^k|\hat{y}^k(t_{ij} + \beta_i;\theta), \sigma_k). \tag{3}$$

Here, β_i is the subject-specific "disease-time" or time-shift at the initial scan, and t_{ij} is the time of scan at visit j, given that $t_{i0} = 0$. We consider the possibility that certain biomarkers, such as a subset of clinical scores, can have a privileged status. Specifically, we set aside some set of biomarkers $\mathbf{c} \in \mathbb{R}^c$ - excluded from $\mathbf{y}$ - that condition individual time-shifts by a simple linear model:

$$P(\beta_i|\mathbf{c}_i) = \prod_j \mathcal{N}(\beta_i + t_{ij}|a + \langle\, \mathbf{b}, \mathbf{c}_{ij}\rangle, \sigma_c), \tag{4}$$

where a and $\mathbf{b}$ are some unknown regression parameters.

2.3 Biomarker Trajectory Conditioning and Likelihood Estimation

To represent the fact that brain function is largely self-contained and locally protected from external inputs, we assume that the number of brain regions affected by $\mathbf{f}$ is small and therefore $\mathbf{f}$ is sparse. Additionally, to prevent exploding timescales which allow all samples to be mapped to flat trajectories, we add an $l2$ penalty to s_k. Thus, our prior on the dynamic system parameters is

$$P(\theta) \propto \exp[-w_f|\mathbf{f}| - w_s s_t^2], \tag{5}$$

Combining the terms in (3),(4),(5), we have the posterior for θ and β:

$$P(\theta, \beta|\mathbf{y}, \mathbf{c}) = \prod_{ijk} P(y_{ij}^k|\theta, \beta_i) \times P(\theta) \times \prod_i P(\beta_i|\mathbf{c}_i) \tag{6}$$

2.4 Model Estimation

We estimate model parameters and subject time-shifts using a generalized Expectation - Maximization approach. We recognize that other optimization strategies may improve the estimation; however, in practice the approach proved robust even for relatively low-signal data such as anatomical imaging in early-stage Parkinson's disease.

E-step: We assume fixed β and optimize θ by minimizing the negative log likelihood of (6). For components of θ that parameterize (2), we numerically integrate both the ODE and the Jacobians of the ODE with respect to f, s_t. In practice, we find that alternating between L-BFGS and Nelder-Mead optimization methods achieves the best results.

M-step: We assume a fixed biomarker trajectory and optimize β. In the case where we assume that $\mathbf{b} \neq \mathbf{0}$, this step is further split into two substeps: 1. estimating global clinical regression parameters $a, \mathbf{b}$ and dispersion σ_c based on current β; 2. optimizing (6) to update β only. The first step is a simple linear problem. The second is a smooth convex problem, solved efficiently with the L-BFGS algorithm.

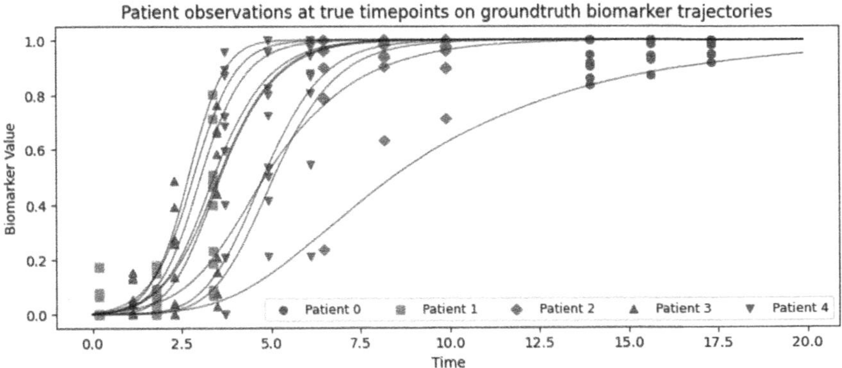

Fig. 1. Five randomly generated patients with their observed biomarker values at the true time. Black lines represent true trajectories.

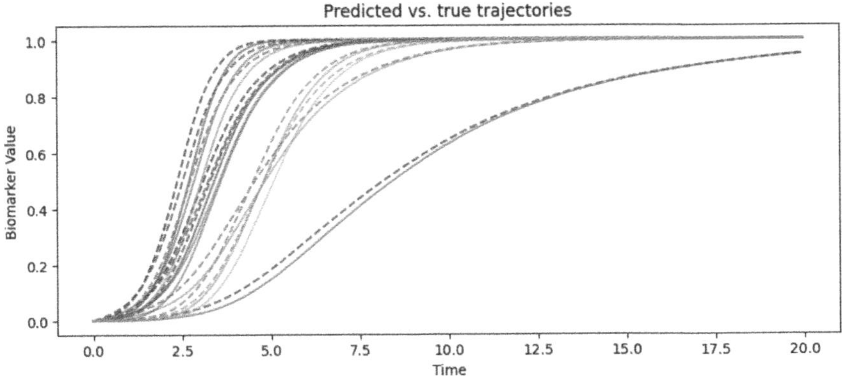

Fig. 2. Fitting synthetic trajectories with the COMIND model. Solid lines are true trajectories, dashed lines are COMIND estimates.

3 Data and Experiments

Synthetic Data Generation. True disease trajectories were generated using (2), with initial conditions equal to zero and **f** being a random realization drawn from Gamma(0,0.05). The connectivity matrix K was generated to mimic the sparsity of the connectome, with the diagonal uniformly set to 0.

After generating the trajectories, 200 patients were simulated with 3 equally spaced consecutive observations per patient starting at a random time-point. The spacing for each patient is randomly distributed on Gamma(2, 0.5). We added Gaussian noise with $\sigma = 0.1$ for each observed value. Each patient had an initial β drawn from a uniform distribution between 0 and $t_{\max}$. We also generated a "privileged" clinical feature. Clinical regression parameters were randomly chosen between [1,5] for slope and [0,10] for the bias term. A value representing a single clinical score was then generated for each observation as follows: cognitive score $= a * (t_{ij} + \mathcal{N}(0, 1)) + b$. Examples of trajectory and sample generation are seen in Fig. 1.

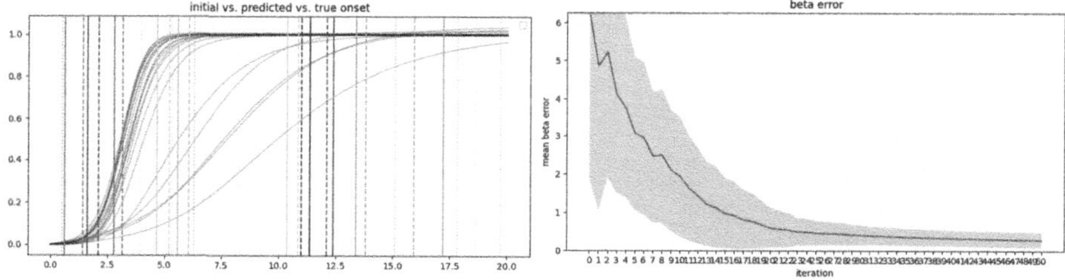

Fig. 3. Fitting synthetic timeshifts (left) and evolution of timeshift prediction error (right) with the COMIND model. Solid lines are true trajectories/timeshifts, faint dotted lines are initial guesses, and dashed lines are final estimates.

3.1 Parkinson's Disease Imaging and Clinical Data

The PPMI study employs a standardized high-resolution 3D T1-weighted volumetric sequence (MP-RAGE/IR-FSPGR) acquired on 3T scanners across 50 international sites (sagittal $1.0 \times 1.0 \times 1.0$ mm isotropic scan, 256^3 FOV). We used 146 PPMI subjects with Parkinson's Disease who had at least one follow-up scan. T1-weighted scans were processed using FreeSurfer 7.0 longitudinal pipeline to harmonize cortical parcellations (Desikan-Killiany atlas, DK) across time points. We use cortical thickness measures in the 68 DK regions as our primary biomarker. There were a total of 504 observations, resulting in 34,272 unique imaging measures.

Additional measures included Montreal Cognitive Assessment (MoCA), Tremor-dominant sub-score (TD-Score) and Postural Instability-Gait Disorder

Subscore (PIGD-score) from MDS-UPDRS-III symptom-specific items, Hoehn-Yahr (HY) stage and Neuronal alpha-Synuclein Disease integrated staging (NSD-ISS). HY and NSD-ISS progress in discrete steps from stage 0/1 to stage 6 (higher=more severe). Lower MoCA and higher TD/PIGD scores imply greater symptom severity.

3.2 Connectome Model

We constructed a canonical connectome from structural connectivity matrices of 794 subjects in the Human Connectome Project. These matrices were generated from preprocessed diffusion MRI data using MRtrix3 [11]. Tractography was performed in an anatomically constrained manner based on T1-weighted MRI. We used constrained spherical deconvolution [12] and intensity normalization to derive an initial set of 40 million tractograhy streamlines (max length = 250, FA cutoff = 0.06). We applied SIFT2 [13] to reconstruct whole-brain streamlines 106. Streamlines were then mapped onto the 68 cortical Desikan-Killiany regions to produce subject-specific connectomes. These were then normalized and averaged. We excluded any self-connections from our connectome, setting the diagonal to zero for imaging-based biomarkers.

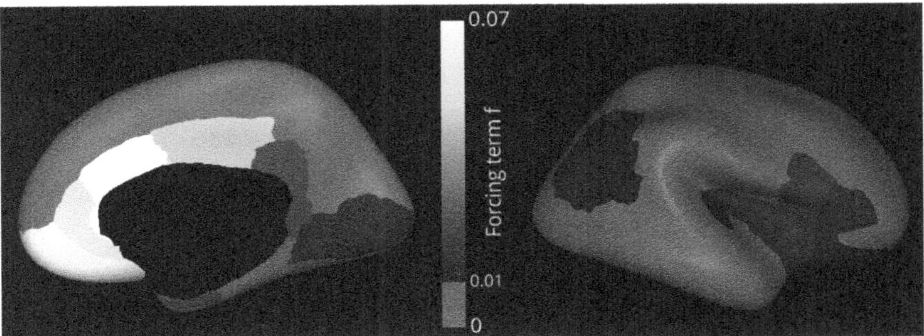

Fig. 4. Map of estimated external neurodegeneration components **f** in the Parkinson's Disease model.

3.3 Synthetic Experiments

We applied our inference algorithm for COMIND to 200 randomly generated samples with fixed hyperparameters w_s, w_f, and $w_c = \frac{\sigma_k}{\sigma_c}$, since σ_k was constant across k and σ_c was known. We evaluated our performance based on trajectory recovery and sample timeshift estimation.

3.4 Parkinson's Disease Experiments

We randomly selected 20% of our subjects as the hold-out validation set. Next, we performed 3-fold grid search cross-validation to find optimal hyperparameters w_s, w_f, w_c. During the evaluation stage, we used the learned $\theta, a, \mathbf{b}$ to estimate β on the inner evaluation set of samples. We selected hyperparameters leading to the smallest least-squares error between predicted trajectories and the evaluation set biomarkers. We used MoCA, TD-score and PIGD-score as the set of privileged biomarkers for conditioning patient timeshifts. To evaluate clinical utility, we estimated Kendall's Tau for ordinal correlation between the estimated disease time and (1) Hoehn-Yahr stage, (2) NDS-ISS stage.

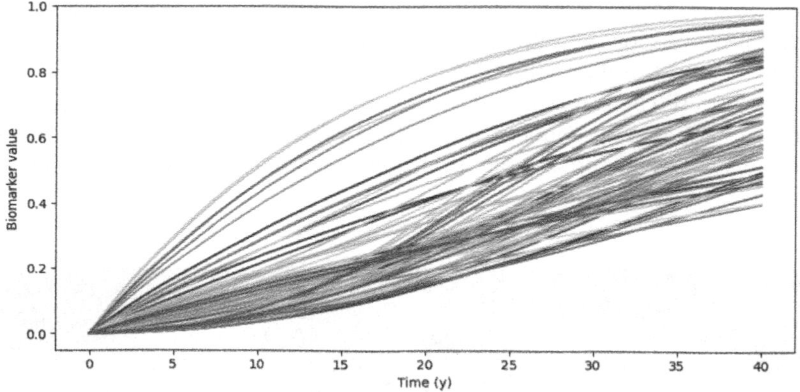

Fig. 5. The estimated normalized trajectories $\mathbf{x}(t)$ from the Desikan-Killiany regions' model of PD over the entire range of time considered.

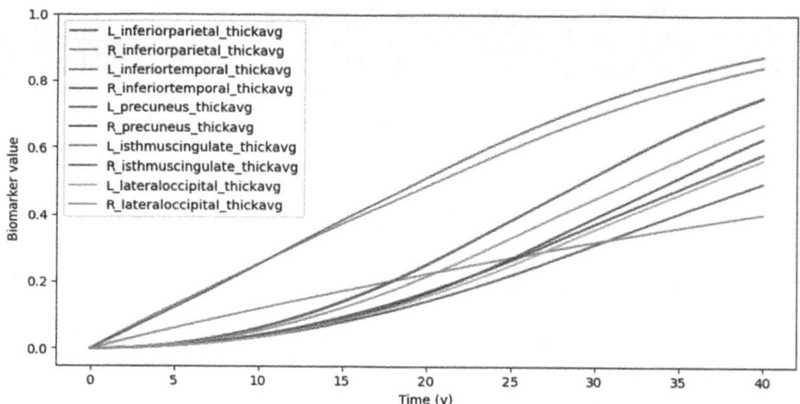

Fig. 6. Selection of trajectories from the 68-region PD model in Fig. 5 for regions showing greatest PD effect in [14]

4 Results

4.1 Synthetic Results

The COMIND model recovered trajectory parameters and sample timeshifts with reasonable accuracy, seen in Figs. 2, 3. Mean β recovery error was 0.3 ± 0.2 on a set of timeshifts drawn uniformly from an imaginary 20-year range. Unsurprisingly, estimating disease time was easier for samples nearer in time to periods of relatively rapid biomarker decline, especially compared to samples at relatively later disease stages.

4.2 Parkinson's Disease Model

The Parkinson's Disease model yielded an external accumulation term **f** that was 45% sparse, and 80% sparse when setting all regions' entries below $0.15 \times$ max to zero. Figure 4 illustrates this; interestingly, the remaining regions are loosely overlapping the standard salience network. Trajectories of the 68 Desikan-Killiany atlas and their subset corresponding to 10 regions most anatomically affected by PD [14] are shown in Figs. 5, 6. Experimentally, the optimization approach of switching between NM and L-BFGS optimizers yielded monotonic decrease in least squares error (LSE) over 20 EM steps, with LSE reduction by a factor of 0.5. Goodness of trajectory fit can be seen in the "spaghetti plots" in Figs. 7, 8. Kendall's Tau for timeshift correlation was $\tau = 0.10$ ($p = 0.19$) for Hoehn-Yahr stage and $\tau = 0.22$ ($p = 0.0053$) for NSD-ISS stage. Distributions for each stage are displayed in Fig. 9.

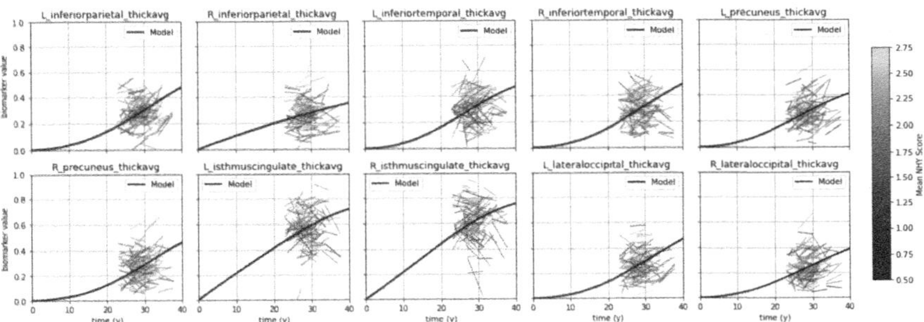

Fig. 7. Training subject trajectories overlaid on the model for 10 selected regions. Color-coding is based on the subject's Hoehn-Yahr stage.

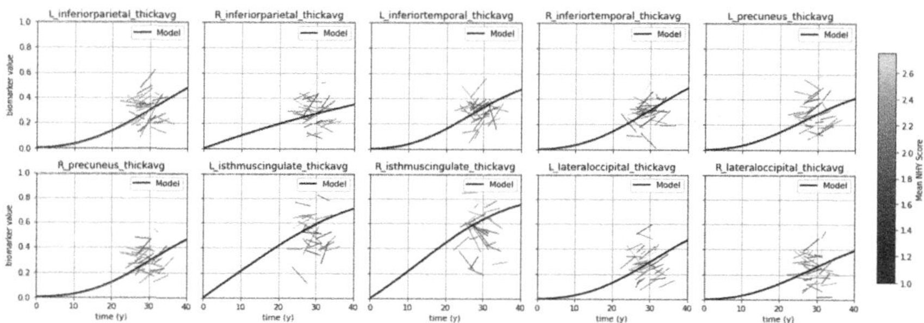

Fig. 8. Validation subject trajectories overlaid on the model for 10 selected regions. Color-coding is based on the subject's Hoehn-Yahr stage.

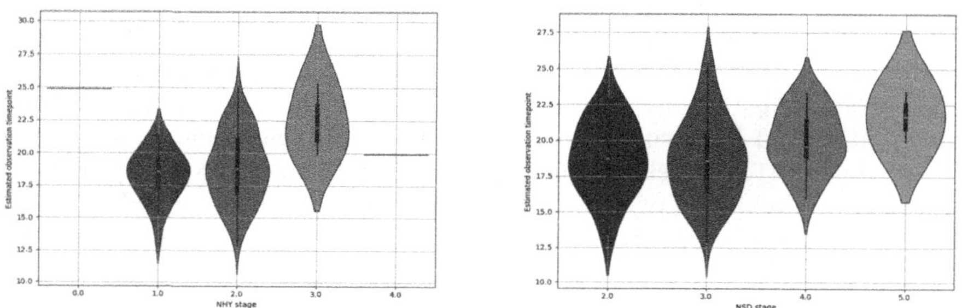

Fig. 9. Distribution of estimated disease-time (timeshift) by Hoehn-Yahr stage (left) and NSD-ISS stage (right). Based on the validation cohort.

5 Conclusion

We have presented COMIND: a progression model of neurodegeneration that combines diffusion-like and logistic accumulation dynamics with the network propagation hypothesis in brain disease. The method's parameter space is made deliberately low-dimensional to enable fitting on relatively small longitudinal MRI cohorts with potentially low disease-related signal. We demonstrate that the approach leads to stable and neurologically plausible results on anatomical MRI data from the PPMI cohort - a notoriously low-signal dataset in which researchers often struggle to find disease effects. It is especially promising that the NSD-ISS staging system appears to be in good agreement with our anatomically-predicted stage; the NSS-IDS is arguably the most comprehensive staging approach in Parkinson's Disease to date, incorporating genetic, DAT, and clinical information to identify patient stage. We also note that some extensions of the method are currently in progress, including the addition of subtyping via the clustering of **f** and adding dynamics to the connectome itself.

Acknowledgments. This work was supported by the Michael J Fox Foundation grant MJFF-021683, Multimodal Dynamic Modeling and Prediction of Parkinsonian Symptom Progression.

References

1. Fonteijn, H.M., et al.: An event-based model for disease progression and its application in familial alzheimer's disease and huntington's disease. Neuroimage 60(3), 1880–9 (2012). https://www.ncbi.nlm.nih.gov/pubmed/22281676
2. Young, A.L., et al.: Uncovering the heterogeneity and temporal complexity of neurodegenerative diseases with subtype and stage inference. Nat. Commun. 9(1), 4273 (2018). https://www.ncbi.nlm.nih.gov/pubmed/30323170
3. Marinescu, R.V., et al.: DIVE: a spatiotemporal progression model of brain pathology in neurodegenerative disorders. Neuroimage 192, 166–177 (2019). https://www.ncbi.nlm.nih.gov/pubmed/30844504
4. Viani, A., Gutman, B.A., d'Angremont, E., Lorenzi, M.: Disease progression modelling and stratification for detecting sub-trajectories in the natural history of pathologies: application to parkinson's disease trajectory modelling. In: Schroder, A., et al. Medical Image Computing and Computer Assisted Intervention – MICCAI 2024 Workshops. MICCAI 2024. LNCS, pp. 3–14. Springer, Cham. https://doi.org/10.1007/978-3-031-84525-3_1
5. Puglisi, L., Alexander, D.C., Ravì, D.: Enhancing Spatiotemporal Disease Progression Models via Latent Diffusion and Prior Knowledge. In: Linguraru, M.G., et al. Proceedings of Medical Image Computing and Computer Assisted Intervention - MICCAI 2024. LNCS, vol. LNCS 15002. Springer, Cham (2024). https://doi.org/10.1007/978-3-031-72069-7_17
6. Raj, A., LoCastro, E., Kuceyeski, A., Tosun, D., Relkin, N., Weiner, M.: Network diffusion model of progression predicts longitudinal patterns of atrophy and metabolism in alzheimer's disease. Cell reports 10(3), 359–369 (2015). https://www.ncbi.nlm.nih.gov/pubmed/25600871https://www.ncbi.nlm.nih.gov/pmc/PMC5747552/
7. Garbarino, S., Lorenzi, M.: Investigating hypotheses of neurodegeneration by learning dynamical systems of protein propagation in the brain. NeuroImage 235, 117980 (2021). https://www.sciencedirect.com/science/article/pii/S1053811921002573
8. Abi Nader, C., Ayache, N., Frisoni, G.B., Robert, P., Lorenzi, M.: Simulating the outcome of amyloid treatments in alzheimer's disease from imaging and clinical data. Brain Commun. 3(2), fcab091 (2021). https://doi.org/10.1093/braincomms/fcab091
9. Young, A.L., et al.: Data-driven modelling of neurodegenerative disease progression: thinking outside the black box. Nat. Rev. Neurosci. **25**(2), 111–130 (2024)
10. Moravveji, S., Doyon, N., Mashreghi, J., Duchesne, S.: A scoping review of mathematical models covering alzheimer's disease progression. Front Neuroinform 18, 1281656 (2024)
11. Tournier, J., et al.: Mrtrix3: a fast, flexible and open software framework for medical image processing and visualisation. NeuroImage 202, 116137 (2019). https://www.sciencedirect.com/science/article/pii/S1053811919307281
12. Tournier, J., Calamante, F., Connelly, A.: Robust determination of the fibre orientation distribution in diffusion MRI: non-negativity constrained super-resolved spherical deconvolution. NeuroImage 35(4), 1459–1472 (2007). https://www.sciencedirect.com/science/article/pii/S1053811907001243

13. Betzel, R., Griffa, A., Hagmann, P., Mišić, B.: Distance-dependent consensus thresholds for generating group-representative structural brain networks. Net. Neuro. 3(2), 475–496 (2019). https://doi.org/10.1162/netn_a_00075
14. Laansma, M.A., et al.: International multicenter analysis of brain structure across clinical stages of parkinson's disease. Mov. Disord. 36(11), 2583–2594 (2021). https://www.ncbi.nlm.nih.gov/pubmed/34288137

CLIP-KOA: Enhancing Knee Osteoarthritis Diagnosis with Multi-modal Learning and Symmetry-Aware Loss Functions

Yejin Jeong and Donghun Lee(✉)

Department of Mathematics, Korea University, Seoul 02841, Republic of Korea
{yejin_mds,holy}@korea.ac.kr

Abstract. Knee osteoarthritis (KOA) is a universal chronic musculoskeletal disorders worldwide, making early diagnosis crucial. Currently, the Kellgren and Lawrence (KL) grading system is widely used to assess KOA severity. However, its high inter-observer variability and subjectivity hinder diagnostic consistency. To address these limitations, automated diagnostic techniques using deep learning have been actively explored in recent years. In this study, we propose a CLIP-based framework (CLIP-KOA) to enhance the consistency and reliability of KOA grade prediction. To achieve this, we introduce a learning approach that integrates image and text information and incorporate Symmetry Loss and Consistency Loss to ensure prediction consistency between the original and flipped images. CLIP-KOA achieves state-of-the-art accuracy of 71.92% on KOA severity prediction task, and ablation studies show that CLIP-KOA has 2.42% improvement in accuracy over the standard CLIP model due to our contribution. This study shows a novel direction for data-driven medical prediction not only to improve reliability of fine-grained diagnosis and but also to explore multimodal methods for medical image analysis. Our code is available at https://github.com/AIML-K/CLIP-KOA.

Keywords: Vision Language Model · Knee Osteoarthritis Classification · Symmetry and Consistency Loss · CLIP

1 Introduction

Osteoarthritis is the leading chronic musculoskeletal disease worldwide, affecting approximately 15% of the population, including around 43 million people in the United States [3,16]. Knee OA, in particular, is a leading cause of disability and is characterized by a progression from intermittent pain during weight-bearing activities to persistent, chronic pain [12]. OA is not merely a degenerative change in articular cartilage but a complex condition that affects the entire joint. Therefore, early diagnosis and tailored treatment strategies are essential for its effective prevention and management.

B. Hou and T. S. Mathai (Eds.): LMID 2025, LNCS 16184, pp. 125–134, 2026.
https://doi.org/10.1007/978-3-032-16128-4_12

One of the most widely used methods for diagnosing OA is the KL grading system, a radiological osteoarthritis assessment framework proposed in 1957 [10]. It evaluates OA severity on a scale from 0 to 4 based on radiological (X-ray) images. However, the KL grading system relies heavily on the physician's subjective judgment, leading to limitations that can result in variability depending on the physician's experience [4]. In fact, the inter-rater reliability of the KL system has been reported to range from 0.54 to 0.79, with adjacent grades, such as KL 1 and KL 2, being particularly prone to misclassification due to their subtle differences [2,19]. To address these limitations, automated OA diagnosis using deep learning has been actively explored, highlighting the growing need for more objective and quantitative evaluation methods.

Existing studies have employed various techniques for OA diagnosis, with a particular focus on the KL grading system. However, this system has significant limitations due to high inter-observer variability and reliance on subjective assessment. To address these issues, automated KOA diagnosis systems based on deep learning and machine learning have been actively explored [21].

Some studies have utilized multi-class classification approaches to predict KL grades [7,9,14,15,17,22,23], while others have applied ordinal regression techniques to account for the ordinal nature of KL grading [24]. In particular, [22] proposed a Siamese CNN-based approach for KL grading, while [14,15] addressed the difficulty of early diagnosis by focusing on texture and shape features. The most recent work [23] leverages Vision Transformers and novel patch-based augmentation strategies, reporting strong binary classification performance for early-stage OA.

Additionally, Medical Vision-Language Pretraining (VLP) has emerged as a key approach to mitigating medical data sparsity. It has been actively explored to enhance performance in a range of medical image analysis tasks [20]. Contrastive Language-Image Pretraining (CLIP) models [18] have demonstrated strong generalization capabilities across various downstream vision tasks. They have gained significant attention for their high classification accuracy, particularly in imbalanced or stylistically diverse datasets. Recent studies have explored the application of CLIP's pretraining techniques to the medical imaging domain, leading to various reported use cases [6]. Notable examples include Diabetic Retinopathy analysis [25], Histopathology image interpretation [8], and PubMedCLIP, which incorporates Pathology, Blood, and Breast imaging data [5], as well as BiomedCLIP [26]. Additionally, research on integrating ordinal relationships between classes in CLIP models has gained traction [11,25], offering promising applications for tasks requiring quantitative assessment in medical imaging.

In this study, we propose a CLIP-based framework (CLIP-KOA) for the automatic and consistent classification of Knee Osteoarthritis (KOA). Given the high inter-observer variability and subjective nature of the existing KL grading system, we introduce a novel learning approach that integrates both image and text modalities to address these limitations. The main contributions of this work are threefold. First, unlike existing KOA grading classification models that rely solely on image data, CLIP-KOA integrates both image and textual information to enable more sophisticated KOA severity prediction. This allows the model to go beyond simple KL grade (04) classification by learning both the meaning and features of each grade, leading to a more reliable KOA assessment. Second, we leverage the inherent symmetry of KOA images to design Symmetry Loss and Consistency Loss, which ensure prediction consistency between the original and flipped images. These loss functions encourage the model to produce consistent KOA ratings regardless of image orientation, thereby improving generalization performance. The schematic illustration of CLIP-KOA is shown in Fig. 1. Third, we empirically demonstrate that CLIP-KOA can predict KOA severity more precisely and consistently than existing CNN-based KOA grading models. While existing KL grading systems use only a fixed scale of 04, CLIP-KOA is designed to leverage textual information to capture more nuanced and subjective descriptions of KOA severity. This work suggests potential future research for overcoming the limitations of current KOA grading prediction methods with more refined multimodal image analysis approaches.

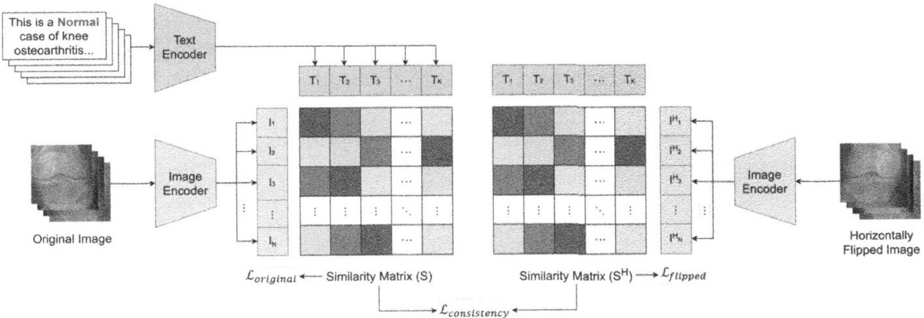

Fig. 1. Symmetry Loss and Consistency Loss framework for KOA severity grading. The original image I and its horizontally flipped version I^H are mapped to similarity matrices $\mathcal{S}$ and $\mathcal{S}^H$ through the image and text encoders. Symmetry Loss is computed using the cross-entropy loss between the similarity matrices and one-hot encoded labels. Jensen-Shannon Divergence (JSD) is used in $\mathcal{L}_{consistency}$ to measure the difference between the distributions of $\mathcal{S}$ and $\mathcal{S}^H$. The final loss function integrates Symmetry Loss and Consistency Loss to enhance the robustness of KOA severity classification.

2 Methods

2.1 Problem Formulation

A KOA image I_i is provided as an input, with its corresponding severity grade G_i. Thus, the dataset can be represented as a collection of paired samples: $\{I_i, G_i\} \in D$ where D is the knee osteoarthritis dataset. The image encoder maps I_i to an image embedding x_i, defined as : $\mathcal{X} = \{x_i\}_{i=1}^{N}$ where N is the total number of images. Given that there are K distinct grades, we define the set of text embeddings as: $\mathcal{T} = \{t_j\}_{j=1}^{K}$. The similarity matrix $\mathcal{S}$ is computed from the set $\mathcal{X}$ and $\mathcal{T}$ using cosine similarity: $\mathcal{S} = [s_{i,j}]_{N\times K}$, $s_{i,j} = x_i \cdot t_j^T$. The objective is to model a mapping function $f_\theta : I_i \rightarrow G_i$, which maps an input knee osteoarthritis image I_i to its corresponding grade G_i. This function is implemented as a deep neural network with learnable parameters θ.

2.2 Loss Function

The key idea of Symmetry Loss and Consistency Loss is that the model should maintain consistent predictions even when the left and right sides of the knee X-ray images are mixed. In other words, if the same knee osteoarthritis image is flipped left and right, the model should predict the same KL grade, and the prediction probability distribution should be similar.

Symmetry Loss. The main loss consists of both the image-text loss and horizontal flip image-text loss. Given an image I_i, its corresponding embedding vector x_i and the text embedding vector t_j, the similarity score $s_{i,j}$ is computed as their inner product. The similarity matrix $\mathcal{S} \in \mathbb{R}^{N\times K}$ represents the pairwise inner products between image and text embeddings. Let I_i^H be the horizontal flipped image of image I_i. Then, $\mathcal{S}^H$ represents the similarity matrix computed for the flipped image-text pairs. The label corresponding to $\mathcal{S}$ is denoted as Y and is formulated as:

$$Y = \mathbb{I}(S) = \begin{cases} 1 & \text{ground-truth grade} \\ 0 & \text{otherwise} \end{cases} . \quad (1)$$

In the general CLIP framework, we use image-text pairs (I_i, T_i) to learn the similarity between images and their corresponding text labels. That is, it is optimized by increasing the similarity between the image and its correctly matched text pair (I_i, T_i) and decreasing the similarity of the incorrect pair (I_i, T_j). However, since the goal of this study is to find labels that correctly match the text representing the corresponding rating of an image, the value of K is set to the number of grades defined in a given problem. Therefore, the labels in Y are represented as one-hot encoding vectors corresponding to their respective grades. Based on these rules, image-text loss and horizontal flipped image-text loss can be expressed as:

$$\mathcal{L}_{\text{symmetry}} = \mathcal{L}_{\text{original}} + \mathcal{L}_{\text{flipped}} = \text{CELoss}(S, Y) + \text{CELoss}(S^H, Y) . \quad (2)$$

Consistency Loss. While preserving the characteristics of a given knee osteoarthritis image I, the similarity score matrix $\mathcal{S}^H$ of the horizontally flipped image I^H and the similarity score matrix $\mathcal{S}$ of the original image should share the same predictive probability distribution. Therefore, Jensen-Shannon Divergence (JSD) [13] is used to quantitatively evaluate the similarity between the two probability distributions Eq. (3).

$$\mathcal{L}_{\text{consistency}} = JSD(P||Q) = \frac{1}{2}D_{KL}(P||M) + \frac{1}{2}D_{KL}(Q||M) \,. \tag{3}$$

We define the predictive probability distribution of the original image as $P = \text{softmax}(\mathcal{S})$ and the predictive probability distribution of the horizontally flipped image as $Q = \text{softmax}(\mathcal{S}^H)$, and set their mean as $M = \frac{1}{2}(P + Q)$. This serves as a quantitative measure of the difference between two probability distributions P and Q by calculating the KL-Divergence based on the mean M of the two distributions. Finally, the total loss function is defined as Eq. (4). Consistency Loss acts as a regularizer, encouraging the model to maintain consistent predictions across flipped images, ensuring that the same KOA image yields the same KL rating when flipped left and right.

$$\mathcal{L}_{\text{total}} = \frac{1}{2}(\mathcal{L}_{\text{original}} + \mathcal{L}_{\text{flipped}}) + \lambda\mathcal{L}_{\text{consistency}} \,, \tag{4}$$

where λ is a hyperparameter, set to 10 by default, which defines relative weight of the consistency loss term $\mathcal{L}_{\text{consistency}}$ to the remaining portions of the loss.

3 Experiments

Dataset. We used the KOA severity grading dataset to evaluate our method [1]. The KOA dataset contains knee X-ray data for both knee joint detection and knee KL grading. The dataset consists of knee radiographs from 4796 participants. All images were resized to 224 $\times$ 224 pixels to ensure consistency in resolution. The processed dataset included a total of 8260 images, which were split into training, validation, and test sets in a 7:1:2 ratio. Table 1 presents the distribution of the dataset across different KL grades. For the text labels, we utilized the grading descriptions provided on Kaggle, which define the severity levels of knee osteoarthritis from Grade 0 to Grade 4. These descriptions serve as the standard criteria for KL grading, outlining the progression of osteoarthritis from healthy knees (Grade 0) to severe cases (Grade 4). Based on these descriptions, we generated text labels corresponding to each KL grade to enhance the interpretability and consistency of the dataset.

Implementation Details. Our model is built upon the ViT-B/16 Transformer architecture, utilizing a pre-trained CLIP backbone for both image and text encoding. The image encoder is based on a Vision Transformer (ViT), while the text encoder employs a masked self-attention Transformer. The implementation was carried out in PyTorch and trained on an NVIDIA RTX A6000 GPU. During training, we employed the AdamW optimizer with a base learning rate

Table 1. The distribution of knee osteoarthritis dataset

	Normal	Doubtful	Minimal	Moderate	Severe	Total
Train	2286	1046	1516	757	173	5578
Validation	328	153	212	106	27	826
Test	639	296	447	223	51	1656
Total	3253	1495	2175	1086	251	8260

Table 2. Comparison of accuracy across previous studies and the proposed CLIP-KOA model on the same dataset.

	Method	Accuracy
Chen et al. (2019) [2]	VGG 19 - Ordinal	69.6%
Feng et al. (2021) [7]	Attention module	70.23%
Yong et al. (2022) [24]	DenseNet 161 - ORM	70.23%
Jain et al. (2024) [9]	OsteoHRNET	71.74%
Ours	**CLIP-KOA**	**71.92%**

of 1×10^{-5} and a weight decay of 1×10^{-6}. The learning rate was adjusted using a OneCycleLR scheduler. We set the batch size to 64 for efficient training. Additionally, the knee joint images were normalized using the mean and standard deviation of the training dataset to ensure consistent input distributions. To assess model performance, we evaluated classification results using accuracy, recall, precision, and F1-score, ensuring a comprehensive analysis of its effectiveness in knee osteoarthritis severity grading.

4 Results

In this study, we evaluated the performance of the proposed CLIP-KOA model in classifying the severity of KOA and compared it with previous studies using the same dataset. As shown in Table 2, the comparison included models such as VGG 19 - Ordinal [2], Attention Module [7], DenseNet 161 - ORM [24], and OsteoHRNET [9]. In the experimental results, the CLIP-KOA model outperformed the existing models, achieving an accuracy of 71.86%, which is 0.12% points higher than the previous best performer, OsteoHRNET (71.74%).

Traditional CNN-based KOA classification models use only images as input and rely on extracting features from limited data. This approach makes it difficult to overcome data limitations, and performance tends to plateau around 70%. In contrast, the CLIP-KOA model applies multimodal learning, incorporating textual data that includes detailed descriptions of grades rather than simply learning the grades themselves. This enables a more sophisticated differentiation of KOA severity.

In particular, the CLIP-KOA model is designed to go beyond the simple classification of KL grades (04) by learning the descriptions associated with each

Table 3. Ablation study on the impact of Symmetry and Consistency losses

	Accuracy (%)	Precison	Recall	F1-Score
CLIP (baseline)	69.50	0.677	0.698	0.680
+ Symmetry Loss	68.66	0.680	0.709	0.692
+ Consistency Loss	70.41	0.694	0.704	0.690
+ Symmetry + Consistency (CLIP-KOA)	71.92	0.732	0.695	0.708

grade, allowing the model to develop a richer understanding of KOA severity. The existing KL Grading system consists of a fixed five-level scale (04), which has the limitation of not fully capturing the nuances of patients' actual symptoms and interpretations. However, CLIP-KOA utilizes textual keys to learn not only the numerical value of each grade but also its meaning and associated features, enabling a more detailed and precise KOA assessment. This approach is expected to contribute to the refinement of KOA grading and the development of a more sophisticated diagnostic system in the future.

Ablation Study. To verify the performance improvement of the CLIP-KOA model, we conducted an ablation study using the existing CLIP model. As shown in Table 3, the CLIP-KOA model achieved a 2.36% increase in accuracy compared to the original CLIP model and demonstrated overall performance improvements in Precision, Recall, and F1-score.

Notably, the discrimination performance for intermediate KOA classes (KL 23) was significantly enhanced, with the prediction accuracy for the Minimal (KL 2) class increasing by 8%, from 0.66 to 0.74, as shown in Fig. 2. This suggests that the CLIP-KOA model effectively refines KOA grade prediction, particularly in distinguishing mild KOA from severe KOA, which is a crucial challenge in KOA classification.

These improvements can be attributed to the introduction of Symmetry Loss and Consistency Loss, which play a crucial role in refining the model's predictive stability and robustness. Symmetry Loss ensures that the model maintains consistency when an image is horizontally flipped, enforcing robustness by aligning the predictions of original and flipped images. Meanwhile, Consistency Loss minimizes the divergence between the probability distributions of the original and flipped images, reinforcing stable predictions across different transformations.

By incorporating these loss functions, the CLIP-KOA model not only enhances classification accuracy but also ensures a more structured and reliable representation of KOA severity. This enables the model to generate more consistent predictions across varying image conditions, ultimately leading to improved differentiation of KOA severity grades and a more robust KOA classification system.

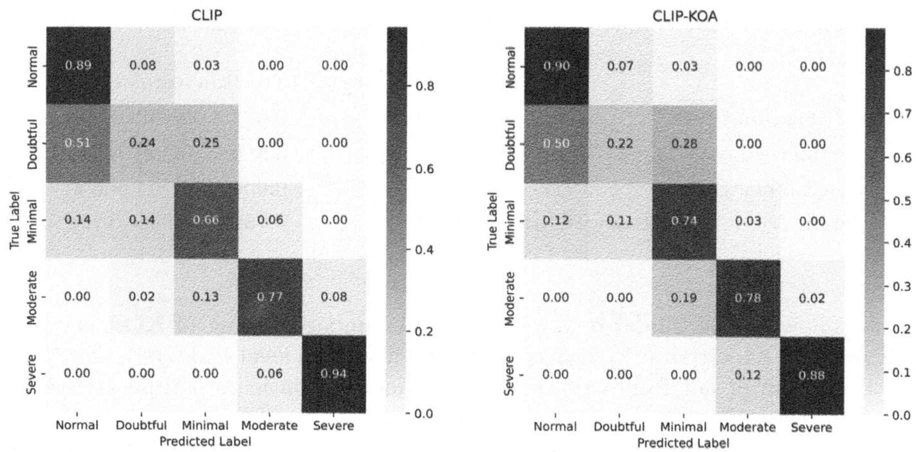

Fig. 2. The figure presents a comparison of confusion matrices for KL grade prediction between the baseline CLIP model (left) and the proposed CLIP-KOA model (right). X-axis (Predicted Label): The five KL grades predicted by the model: Normal, Doubtful, Minimal, Moderate, and Severe. Y-axis (True Label): The actual KL grades assigned in the dataset.

5 Conclusion

We propose a novel method for predicting KL grades from KOA image data. Unlike previous studies that have primarily treated KL grade prediction as a regression or classification problem using only image-based models, our approach incorporates linguistic information to enhance prediction accuracy and consistency. Specifically, we leverage the structural symmetry of KOA images to ensure that original and flipped images yield consistent predictions, and introduce a multi-modal learning strategy that integrates textual descriptions related to KL grading.

The results of this study suggest that the CLIP-KOA model achieves competitive performance compared to existing CNN-based models in KOA severity classification. Unlike traditional models that rely solely on feature learning from image data, CLIP-KOA enhances KOA prediction accuracy and consistency by integrating a language model and introducing a new loss function. This multi-modal approach enables a more structured and robust representation of KOA severity, particularly improving the discrimination of intermediate KL grades (KL 23).

Despite these advancements, challenges remain in the accurate prediction of early-stage KOA, particularly for the Doubtful (KL 1) rating, which was prone to misclassification. To further refine KOA severity classification, future research should explore more diverse textual inputs that provide richer descriptions of KOA symptoms and severity levels. Additionally, optimizing the loss function to incorporate adaptive weighting for specific KL grades could improve prediction accuracy, particularly in the early stages of KOA. Expanding the multimodal learning framework to integrate sequential patient data and clinical text reports may further enhance diagnostic reliability and model generalization.

Disclosure of Interests. The authors have no competing interests to declare that are relevant to the content of this article.

References

1. Chen, P.: Knee osteoarthritis severity grading dataset. Mendeley Data 1(10.17632) (2018)
2. Chen, P., Gao, L., Shi, X., Allen, K., Yang, L.: Fully automatic knee osteoarthritis severity grading using deep neural networks with a novel ordinal loss. Comput. Med. Imaging Graph. **75**, 84–92 (2019) (2019)
3. Cui, A., Li, H., Wang, D., Zhong, J., Chen, Y., Lu, H.: Global, regional prevalence, incidence and risk factors of knee osteoarthritis in population-based studies. EClin. Med. 29 (2020)
4. Culvenor, A.G., Engen, C.N., Øiestad, B.E., Engebretsen, L., Risberg, M.A.: Defining the presence of radiographic knee osteoarthritis: a comparison between the kellgren and lawrence system and oarsi atlas criteria. Knee Surg. Sports Traumatol. Arthrosc. **23**, 3532–3539 (2015)
5. Dao, H.N., Nguyen, T., Mugisha, C., Paik, I.: A multimodal transfer learning approach using PubMedCLIP for medical image classification. IEEE Access (2024)
6. Felfeliyan, B., et al.: Application of vision-language models for assessing osteoarthritis. In: 2024 IEEE International Symposium on Biomedical Imaging (ISBI), pp. 1–4. IEEE (2024)
7. Feng, Y., Liu, J., Zhang, H., Qiu, D.: Automated grading of knee osteoarthritis x-ray images based on attention mechanism. In: 2021 IEEE International Conference on Bioinformatics and Biomedicine (BIBM), pp. 1927–1932 (2021). https://doi.org/10.1109/BIBM52615.2021.9669623
8. Ikezogwo, W., et al.: Quilt-1M: one million image-text pairs for histopathology. Adv. Neural Inf. Process. Syst. **36** (2024)
9. Jain, R.K., Sharma, P.K., Gaj, S., Sur, A., Ghosh, P.: Knee osteoarthritis severity prediction using an attentive multi-scale deep convolutional neural network. Multimedia Tools Appl. **83**(3), 6925–6942 (2024)
10. Kellgren, J.H., Lawrence, J., et al.: Radiological assessment of osteo-arthrosis. Ann. Rheum. Dis. **16**(4), 494–502 (1957)
11. Li, W., et al.: OrdinalCLIP: learning rank prompts for language-guided ordinal regression. Adv. Neural. Inf. Process. Syst. **35**, 35313–35325 (2022)
12. Man, G., Mologhianu, G.: Osteoarthritis pathogenesis-a complex process that involves the entire joint. J. Med. Life **7**(1), 37 (2014)
13. Menéndez, M.L., Pardo, J.A., Pardo, L., Pardo, M.D.C.: The jensen-shannon divergence. J. Franklin Inst. **334**(2), 307–318 (1997)
14. Nasser, Y., El Hassouni, M., Hans, D., Jennane, R.: A discriminative shape-texture convolutional neural network for early diagnosis of knee osteoarthritis from x-ray images. Phys. Eng. Sci. Med. **46**(2), 827–837 (2023)
15. Nasser, Y., Jennane, R., Chetouani, A., Lespessailles, E., El Hassouni, M.: Discriminative regularized auto-encoder for early detection of knee osteoarthritis: data from the osteoarthritis initiative. IEEE Trans. Med. Imaging **39**(9), 2976–2984 (2020)

16. Neogi, T.: The epidemiology and impact of pain in osteoarthritis. Osteoarthritis Cartilage **21**(9), 1145–1153 (2013)
17. Pi, S.W., Lee, B.D., Lee, M.S., Lee, H.J.: Ensemble deep-learning networks for automated osteoarthritis grading in knee x-ray images. Sci. Rep. **13**(1), 22887 (2023)
18. Radford, A., et al.: Learning transferable visual models from natural language supervision. In: International conference on machine learning, pp. 8748–8763. PMLR (2021)
19. Shamir, L., et al.: Knee x-ray image analysis method for automated detection of osteoarthritis. IEEE Trans. Biomed. Eng. **56**(2), 407–415 (2008)
20. Shrestha, P., Amgain, S., Khanal, B., Linte, C.A., Bhattarai, B.: Medical vision language pretraining: a survey (2023). arXiv preprint arXiv:2312.06224
21. Tariq, T., Suhail, Z., Nawaz, Z.: A review for automated classification of knee osteoarthritis using kl grading scheme for x-rays. Biomed. Eng. Lett. **15**(1), 1–35 (2025)
22. Tiulpin, A., Thevenot, J., Rahtu, E., Lehenkari, P., Saarakkala, S.: Automatic knee osteoarthritis diagnosis from plain radiographs: a deep learning-based approach. Sci. Rep. **8**(1), 1727 (2018)
23. Wang, Z., Chetouani, A., Jarraya, M., Hans, D., Jennane, R.: Transformer with selective shuffled position embedding and key-patch exchange strategy for early detection of knee osteoarthritis. Expert Syst. Appl. **255**, 124614 (2024)
24. Yong, C.W., et al.: Knee osteoarthritis severity classification with ordinal regression module. Multimedia Tools Appl. pp. 1–13 (2022)
25. Yu, Q., et al.: CLIP-DR: textual knowledge-guided diabetic retinopathy grading with ranking-aware Prompting. In: Linguraru, M.G., et al. Proceedings of Medical Image Computing and Computer Assisted Intervention – MICCAI 2024. LNCS vol. 15001. Springer, Cham (2024). https://doi.org/10.1007/978-3-031-72378-0_62
26. Zhang, S., et al.: A multimodal biomedical foundation model trained from fifteen million image–text pairs. NEJM AI **2**(1), AIoa2400640 (2025)

Author Index

B. Hou and T. S. Mathai (Eds.): LMID 2025, LNCS 16184, pp. 135–136, 2026.
https://doi.org/10.1007/978-3-032-16128-4